T0210711

SpringerBriefs in Plant Science

SpringerBriefs present concise summaries of cutting-edge research and practical applications across a wide spectrum of fields. Featuring compact volumes of 50 to 125 pages, the series covers a range of content from professional to academic. Typical topics might include:

- A timely report of state-of-the art analytical techniques
- A bridge between new research results, as published in journal articles, and a contextual literature review
- A snapshot of a hot or emerging topic
- An in-depth case study or clinical example
- A presentation of core concepts that students must understand in order to make independent contributions

SpringerBriefs in Plant Sciences showcase emerging theory, original research, review material and practical application in plant genetics and genomics, agronomy, forestry, plant breeding and biotechnology, botany, and related fields, from a global author community. Briefs are characterized by fast, global electronic dissemination, standard publishing contracts, standardized manuscript preparation and formatting guidelines, and expedited production schedules.

Kaiser Iqbal Wani • Tariq Aftab

Plant Molecular Farming

Applications and New Directions

 Springer

Kaiser Iqbal Wani
Department of Botany
Aligarh Muslim University
Aligarh, India

Tariq Aftab ⓘ
Department of Botany
Aligarh Muslim University
Aligarh, India

ISSN 2192-1229   ISSN 2192-1210   (electronic)
SpringerBriefs in Plant Science
ISBN 978-3-031-12793-9   ISBN 978-3-031-12794-6   (eBook)
https://doi.org/10.1007/978-3-031-12794-6

This Springer imprint is published by the registered company Springer Nature Switzerland AG
The registered company address is: Gewerbestrasse 11, 6330 Cham, Switzerland

Preface

Molecular farming is a biotechnological approach that includes the genetic adjustment of agricultural products to create proteins and chemicals for profitable and pharmaceutical purposes. Plant molecular farming describes the manufacture of recombinant proteins and other biologically active product in plants. This approach depends on a genetic transformation of plants that can be accomplished by the methods of stable gene transfer, such as gene transfer to nuclei and chloroplasts, and unstable transfer methods, such as viral vectors. The requirement for recombinant proteins in terms of quality, quantity, and diversity is increasing gradually, which is appealing global attentiveness for the development of new recombinant protein construction technologies and the engineering of orthodox expression systems based on bacteria or mammalian cell cultures. An enormous majority of developing countries cannot afford the high costs of medical treatments resulting from the existing methods. Hence, we need to produce not only the new drugs but also the cheaper versions of those already present in the market. Plant molecular farming is considered a cost-effective technology that has grown and advanced tremendously over the past two decades.

Summarizing the research over the past decades, we bring forth a new book, *Plant Molecular Farming: Applications and New Directions*. We are hopeful that this volume will furnish the requisite of all those who are working or have an interest in the proposed topic.

Aligarh, India   Kaiser Iqbal Wani
                                                                          Tariq Aftab

Contents

Chapter 1
Molecular Farming in Plants: Introduction and Applications

1 Introduction

Plants have been employed as a source of therapeutic products since the dawn of time, and many of today's pharmaceuticals are produced directly or indirectly from them. For example, *Artemisia annua*, a medicinal plant containing bioactive compounds with pleiotropic biological properties, is one of the important sources of artemisinin, a potent anti-malarial drug (Uckun et al., 2021). The recent advancements in genetic engineering and molecular biology have created opportunities to produce a range of plant-based products of medicinal importance. The main outcome of genetic engineering and agricultural biotechnology is the creation of transgenic crops where one or more genes of interest have been incorporated from different sources by different means. The particle gun approach (biolistic method) or *Agrobacterium tumefaciens*-mediated transformation method is being used to develop the majority of transgenic plants. The aim of inserting transgenes into the plant genome is to make it more productive and useful. Even though there have been some controversies related to genetically modified plants, their global planting area is constantly increasing (Turnbull et al., 2021). Currently, farmers grow approximately 190 million hectares of transgenic crops, which is roughly the same size as Mexico's whole surface area (ISAAA, 2020). The conventional usage of genetically modified plants has recently moved beyond food and feed production, with new generations of transgenic plants having unique applications including agricultural-scale pharmaceutical biosynthesis (Paul et al., 2015).

Plants produce a lot of biomass and the protein production in plants can be altered by employing stably transformed plants grown in fields or their suspension cultures in fermenters (Kamenarova et al., 2005). Several different complex mammalian proteins can be produced in transformed plant cell lines and cell suspension cultures. Plants are also capable of producing functioning pharmaceutical proteins due to their strong resemblance to mammalian counterparts. Transgenic plants can

K. I. Wani, T. Aftab, *Plant Molecular Farming*, SpringerBriefs in Plant Science, https://doi.org/10.1007/978-3-031-12794-6_1

also produce recombinant protein-rich organs making them viable options for the long-term storage of such organs (Ma & Wang, 2012). These findings demonstrate the use of genetically modified plants as bioreactors for the manufacture of recombinant proteins to be used as therapeutic and diagnostic tools in the health care sector and life sciences.

2   Different Generations of Genetically Modified Plants

Plant molecular farming is the third generation of agricultural biotechnologies, with the first generation being related to the manipulation of plants for herbicide, pest, and insect resistance. Such plants benefit by reducing the input of agrochemicals and thus costs. One such example of the first generation of genetically modified plants is the introduction of a toxin-producing gene from *Bacillus thermogenesis* in different crops like cotton, brinjal, and corn. When an insect feeds on such plants, Cry protein reaches their gut and is partially degraded releasing the potentially toxic part of the protein which binds with its receptor present on the cells lining the gut of larvae (Ferré et al., 2008; Tabashnik et al., 2008). Subsequent events lead to the bursting of the gut wall of larvae, leading to their death. However, it is not harmful to humans.

The second generation was related to the modification of plants for improving their nutritional and agronomic characteristics or bioremediation (Huot, 2003; Sakpirom et al., 2017). The benefit of the second generation of genetically modified plants is more to consumers, rather than growers. One of the best examples of such a plant is Golden rice, characterized by the presence of high levels of β-carotene in its endosperm. This β-carotene acts as the precursor of fat-soluble vitamin A (retinol) and thus improves the nutritional quality (Beyer et al., 2002).

The third and the most recent generation of genetically modified plants were designed to produce special substances such as chemicals used for industrial purposes or plant-based pharmaceuticals. This technology which uses transgenic plants as protein factories is known as molecular farming.

3   Molecular Farming: Definitions and Applications

The use of genetically engineered plants to produce therapeutic proteins has been referred to by a variety of labels. The term 'pharming' is sometimes used to refer to the use of plants in pharmaceutical synthesis, although it is more commonly used to refer to the use of animals in drug synthesis (Norris, 2005). To avoid this confusion, the manufacture of recombinant pharmaceutical proteins by genetically engineered plants is known as Plant Molecular Farming or Molecular Pharming (Hassan et al., 2011; Price, 2003). Moreover, plant molecular farming is also employed for the commercial synthesis of non-pharmaceutical substances such as cosmetic ingredients (Holtz et al., 2015). Growth factors, monoclonal antibodies, vaccines, enzymes, and cytokines are among the biopharmaceutical molecules that may be produced using this method. Although such products are now produced on a small to medium scale, the maximum potential of plant molecular farming will only be realized if large-scale manufacturing is achieved (Greenham & Altosaar, 2013). It is a promising and new biotechnological approach involving the genetic modification of plants in order to produce recombinant proteins.

According to the Canadian Food Inspection Agency, Plant Molecular Farming is defined as "the use of plants in agriculture to produce biomolecules instead of food, feed, and fiber. Plants with introduced novel traits that produce scientifically, medically or industrially interesting biomolecules are grown as crops and harvested for the biomolecules." The products derived from the plants used in plant molecular farming are divided into two categories as shown in Fig. 1.1.

1. *Primary products*: Antigens (vaccines), antibodies, antibody fragments, structural proteins, drugs, enzymes (therapeutic, industrial, cosmetic, diagnostic), enzyme inhibitors, and therapeutic agents.
2. *By-products*: Secondary metabolites (glucosinolates, phenolic compounds, sugars, starches, tannin, alkaloids, perfumes, aromas, and scents), bioplastics, vitamins, nutraceuticals, cofactors, and fibers.

Due to their complexity and diversity, proteins have wide applications, with over 300 protein-based drugs being approved in the United States and Europe (Walsh, 2018). Due to the wide application of proteins, the requirement for recombinant proteins is fairly strong, with their market value expected to reach US$2.850 billion by 2022, with biopharmaceuticals such as vaccines, antibodies, cytokines, enzymes, growth factors, and enzymes accounting for half of this market, followed by industrial proteins and research reagents (Markets and Markets, 2017).

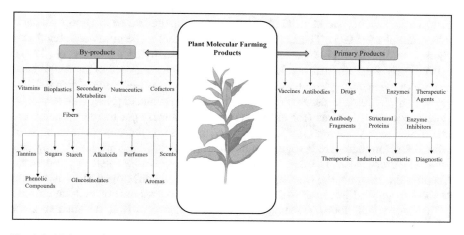

Fig. 1.1 Major products derived from transgenic plants in molecular farming

Since the approval of the first recombinant protein drug, recombinant human insulin named Humulin, by the US FDA in 1982, the recombinant protein industry has grown rapidly (Pham, 2018). The wide usage of therapeutic proteins, particularly protein and peptide drugs is due to their selectivity, excellent potency, and low incidences of toxicity (Ma & Wang, 2012). The majority of clinically accessible protein and peptide therapeutics are produced by using conventional culture-based expression systems involving *E. coli*, mammalian cells, insect, and yeast cells which have proven quite useful. However, these expression systems involve high production costs with low yield and efficacy, with mammalian cells posing concerns about product safety as well. The mammalian cell cultures also have stringent regulatory approvals, hindering the industrial acceptance of such systems.

The increased demand for safe biopharmaceuticals has driven interest in employing plant bioreactors to produce recombinant therapeutic proteins. The improvements in plant molecular farming during the last decade have made plants commercially attractive production systems for the fast production of valuable proteins (Sainsbury & Lomonossoff, 2014; Schillberg et al., 2019). Since the 1990s, most researchers have preferred plants that had already been utilized in earlier experiments since gene transfer procedures were already accessible. This aspiration for the production of recombinant proteins resulted in the creation of a wide range of production techniques, including cell and tissue systems (cell suspension and hairy root cultures), whole plants, and various expression approaches (transient expression systems, stably transformed transplastomic and transgenic plants, diverse protein targeting strategies, and, inducible expression) (Schillberg et al., 2019; Spiegel et al., 2018; Twyman et al., 2005).

The first pharmacologically important protein generated in plants was the human growth hormone, and it was produced in transgenic tobacco and sunflower callus in 1986 (Barta et al., 1986). In 1989, tobacco was used to produce the first plant-based antibody, demonstrating plants' capacity to synthesize complex glycans (Hiatt et al., 1989). A surface antigen of the hepatitis B virus was the first experimental vaccination described from plants (Mason et al., 1992), whereas the first dietary plant-based vaccine was *E. coli* heat-labile enterotoxin generated in transgenic potatoes (Haq et al., 1995). The first plant-derived protein to be commercialized was chicken avidin, a non-pharmaceutical protein produced in maize (Hood et al., 1997).

The early work on plant-made pharmaceuticals was restricted to the use of food crops like maize and rice to build a more efficient and cost-effective vaccine delivery system. In 2002, transgenic maize seeds containing capsid proteins of transmissible gastroenteritis virus (for experimental pig vaccines) got mixed up and were later found in soybean and maize harvests. Concerns about cross-contamination of other field crops hampered progress in this sector, which shifted this interest towards non-food crops such as tobacco, in isolated fields or controlled environments. Keeping this in mind, the US Department of Agriculture developed rules on the use of food crops for Molecular Farming to prevent the occurrence of such an episode in the future (NN, 2002). During 1990s, Dr. J. Christopher Hall, a Canadian plant scientist recognized the advantages of transgenic tobacco in the production of antibody medicines. From 2002 to 2014, Christopher Hall held the Canada Research

Chair in recombinant antibody technology at the University of Guelph where his lab developed a tobacco-plant-based pharmaceutical platform. TrypZean™, a *Zea mays*-derived bovine trypsin intended for commercial applications such as biopharmaceutical processing was the first product commercialized for a broad market (Woodard et al., 2003). A major milestone was achieved in 2006 when the USDA authorized the first plant-derived vaccine against Newcastle disease in chicken. In the same year, Cuba scientists commercialized the first plant-derived monoclonal antibody that identifies hepatitis B virus particles and is used to purify hepatitis B vaccinations (Pujol et al., 2005).

The approval for the human use of the first plant-based pharmaceutical Elelyso (taliglucerase alfa) in 2012, an enzyme generated in carrot cells by Israel-based Protalix Biotherapeutics Inc. for the treatment of Gaucher's disease provided an impetus for the further exploration of the healing potential of plants as well as their commercial non-pharmaceutical applications. The FDA approved LeafBio Inc. (Mapp Biopharmaceutical, Inc) Fast Track status in 2015 for ZMapp, a plant-based medication designed to treat Ebola virus sickness (https://www.plantformcorp.com/history-of-biopharming.aspx; Last accessed April 14, 2022). Some of the major milestones that have been achieved in plant molecular farming have been shown in Fig. 1.2. The use of genetically engineered plant cell systems and plants to serve as drug-producing factories including protein-based pharmaceuticals, vaccines, and

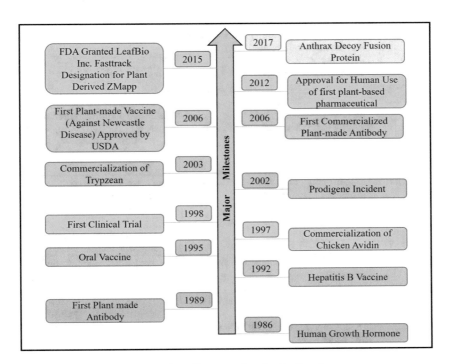

Fig 1.2 A brief history showing major milestones of molecular Farming

antibodies, collectively called plant-made therapeutics has an important role to play in modern health care.

Plants offer several benefits over other expression platforms, especially when the target proteins are difficult to manufacture in traditional systems, require specific qualitative features such as specific glycan profiles, or are in high demand. Plant-based systems also don't support mammalian viral multiplication, which meets customer demand for certified animal-free goods (Spiegel et al., 2018). The target protein is extracted and refined or employed as part of a crude extract in plant molecular farming, with plants serving as hosts that are normally killed or discarded at the end (Buyel, 2019). The comparison of advantages and disadvantages between different expression platforms is summarized in Table 1.1.

Taking into consideration the desired products, the major thrust area of research is related to biopharmaceuticals which have greater added value than technical and diagnostic proteins. In this context, three major classes of protein products have

Table 1.1 Comparison of different expression systems used for the generation of recombinant proteins

Expression System	Advantages	Disadvantages
Bacterial cells	Low cost, easy to manipulate with high expression of proteins Their regulatory protocols are established	Accumulation of endotoxins Lack of post-translational modifications Improper folding
Yeast cells	Easy to manipulate and quick growth Media and culture requirements are inexpensive Presence of post-translational modifications	Tough cell walls make cell disruption difficult Glycosylation potential is limited
Insect cells	The high expression rate of recombinant proteins Ability to produce complex proteins Presence of post-translational modifications Proper folding of proteins	Expensive and time taking process Media and culture requirements are expensive
Mammalian cells	Proper folding of expressed proteins Presence of post-translational modifications Regulatory approvals are already established	High manufacturing costs Media and cultural requirements are both costly and difficult to meet Concerns about product safety Stringent regulatory approvals Hindrance in industrial acceptance
Plants	Large scale production and easily affordable Presence of post-translational modifications almost like mammalian systems Do not support the growth of human pathogens Can support consumer demands for animal-free products	Glycosylation capacity is limited They have regulatory compliance

evolved: antibodies, vaccines, and replacement human proteins such as blood products (human serum albumin), replacement proteins for illness (e.g., glucocerebrosidase for Gaucher's disease, insulin for diabetes), or cytokines and growth factors (Spiegel et al., 2018). The most widely expressed proteins in plants include recombinant antibodies, antibody fusion proteins, and antibody fragments as they are stable, easy to characterize, and important in pharmaceuticals (Vasilev et al., 2016; Walsh, 2018). Their purification is simple, and their functioning may be determined with simple binding tests. Furthermore, they can also build up to large concentrations (>100 mg/L culture media or >100 mg/kg fresh plant weight). However, in terms of the antibody yield, plants lag behind Chinese Hamster Ovary (CHO) cells making their commercial utility uncertain for the time being. However, there might be some niche markets that need to be identified where plants may hold an edge over CHO cells or other platforms (Schillberg et al., 2019).

The recent advancement in plant transgenic technology has also revealed particular platforms and regulatory standards allowing for the production of unique goods in plants rather than bacterial or mammalian cells (Schillberg et al., 2013; Twyman et al., 2003). Protein extraction from field-grown transgenic plants might be a feasible vaccine manufacturing approach, combining plant biology with medical science innovations. Non-polluting and carbon-neutral enzymatic biodiesel fuel products are one of the most profitable applications of biotechnology-derived enzymes (Fischer et al., 1999). Additionally, plants might possibly be more cost-effective than other standard platforms for the expression of polymer breakdown enzymes. The enzymes could be purified, used as crude extracts, or directly expressed in biofuel crops (Ma et al., 2005). Furthermore, the biorefinery idea of plant molecular farming will make it easier to develop low-waste processes in which practically all of the raw material is turned into valuable products (Schillberg et al., 2013). The growth of paper production, biofuel, and feed/food additive sectors may be more engaged in plant molecular farming since non-pharmaceutical antibody generation in various forms has been established for applications in food processing, diagnostics, and quality assurance (Tschofen et al., 2016).

References

Barta, A., Sommergruber, K., Thompson, D., Hartmuth, K., Matzke, M. A., & Matzke, A. J. M. (1986). The expression of a nopaline synthase human growth hormone chimeric gene in transformed tobacco and sunflower callus tissue. *Plant Molecular Biology, 6*, 347–357.

Beyer, P., Al-Babili, S., Ye, X., Lucca, P., Schaub, P., Welsch, R., & Potrykus, I. (2002). Golden rice: Introducing the β-carotene biosynthesis pathway into rice endosperm by genetic engineering to defeat vitamin A deficiency. *The Journal of Nutrition, 132*(3), 506S–510S.

Buyel, J. F. (2019). Plant molecular farming–integration and exploitation of side streams to achieve sustainable biomanufacturing. *Frontiers in Plant Science, 9*, 1893.

Ferré, J., Rie, J. V., & MacIntosh, S. C. (2008). Insecticidal genetically modified crops and insect resistance management (IRM). In *Integration of insect-resistant genetically modified crops within IPM programs* (pp. 41–85). Springer.

Fischer, R., Emans, N., Schuster, F., Hellwig, S., & Drossard, J. (1999). Towards molecular farming in the future: using plant-cell-suspension cultures as bioreactors. *Biotechnology and Applied Biochemistry, 30*(2), 109–112.

Greenham, T., & Altosaar, I. (2013). Molecular strategies to engineer transgenic rice seed compartments for large-scale production of plant-made pharmaceuticals. *Rice Protocols, 956,* 311–326.

Haq, T. A., Mason, H. S., Clements, J. D., & Arntzen, C. J. (1995). Oral immunization with a recombinant bacterial antigen produced in transgenic plants. *Science, 268,* 714.

Hassan, S. W., Waheed, M. T., & Lössl, A. G. (2011). New areas of plant-made pharmaceuticals. *Expert Review of Vaccines, 10*(2), 151–153.

Hiatt, A., Caffferkey, R., & Bowdish, K. (1989). Production of antibodies in transgenic plants. *Nature, 342,* 76–78.

Holtz, B. R., Berquist, B. R., Bennett, L. D., Kommineni, V. J., Munigunti, R. K., White, E. L., & Marcel, S. (2015). Commercial-scale biotherapeutics manufacturing facility for plant-made pharmaceuticals. *Plant Biotechnology Journal, 13*(8), 1180–1190.

Hood, E. E., Witcher, D. R., Maddock, S., Meyer, T., Baszczynski, C., Bailey, M., Flynn, P., Register, J., Marshall, L., Bond, D., Kulisek, E., Kusnadi, A., Evangelista, R., Nikolov, Z., Wooge, C., Mehigh, R. J., Hernan, R., Kappel, W. K., Ritland, D., et al. (1997). Commercial production of avidin from transgenic maize: Characterization of transformant, production, processing, extraction and purification. *Molecular Breeding, 3,* 291–306.

Huot, M. F. (2003). Plant molecular farming: Issues and challenges for Canadian regulators. *Option Consommateurs.* Available online at: http://www.ic.gc.ca/app/oca/crd/dcmnt.do?lang=eng&id=1597 (Last accessed April 10, 2022).

ISAAA. (2020). ISAAA brief 55–2019: Executive summary. Available Online at: https://www.isaaa.org/resources/publications/briefs/55/executivesummary/default.asp (Last accessed April 14, 2022).

Kamenarova, K., Abumhadi, N., Gecheff, K., & Atanassov, A. (2005). Molecular farming in plants: An approach of agricultural biotechnology. *Journal of Cell and Molecular Biology, 4*(4), 77–86.

Ma, S., & Wang, A. (2012). Molecular farming in plants: An overview. *Molecular Farming in Plants: Recent Advances and Future Prospects,* 1–20. https://doi.org/10.1007/978-94-007-2217-0_1

Ma, J. K. C., Barros, E., Bock, R., Christou, P., Dale, P. J., Dix, P. J., & Twyman, R. M. (2005). Molecular farming for new drugs and vaccines: current perspectives on the production of pharmaceuticals in transgenic plants. *EMBO Reports, 6*(7), 593–599.

Markets and Markets. (2017). Protein expression market by type (*E. coli,* mammalian, yeast, Pichia, insect, baculovirus, cell-free), products (competent cells, reagents, instruments, services), application (therapeutic, research, industrial) & end user - global forecast to 2022. *Report Code BT,* 2435.

Mason, H. S., Lam, D. M., & Arntzen, C. J. (1992). Expression of hepatitis B surface antigen in transgenic plants. *Proceedings of the National Academy of Sciences of the United States of America, 89,* 11745.

NN. (2002). Guidance for Industry. Drugs, biologics, and medical devices derived from bioengineerd plants for use in humans and animals.

Norris, S. (2005). Molecular farming. *Parliamentary Information and Research Service: PRB,* 05–09E.

Paul, M. J., Thangaraj, H., & Ma, J. K. C. (2015). Commercialization of new biotechnology: a systematic review of 16 commercial case studies in a novel manufacturing sector. *Plant Biotechnology Journal, 13*(8), 1209–1220.

Pham, P. V. (2018). Medical biotechnology: Techniques and applications. In *Omics technologies and bio-engineering* (pp. 449–469). Academic Press.

Price, B. (2003). Conference on plant-made pharmaceuticals. 16-19 march 2003, Quebec City, Quebec, Canada. *IDrugs, 6*(5), 442–445.

Pujol, M., Ramirez, N. I., Ayala, M., Gavilondo, J. V., Valdes, R., Rodriguez, M., Brito, J., Padilla, S., Gomez, L., & Reyes, B. (2005). An integral approach towards a practical application for a plant-made monoclonal antibody in vaccine purification. *Vaccine, 23*, 1833–1837.

Sainsbury, F., & Lomonossoff, G. P. (2014). Transient expressions of synthetic biology in plants. *Current Opinion in Plant Biology, 19*, 1–7.

Sakpirom, J., Kantachote, D., Nunkaew, T., & Khan, E. (2017). Characterizations of purple non-sulfur bacteria isolated from paddy fields, and identification of strains with potential for plant growth-promotion, greenhouse gas mitigation and heavy metal bioremediation. *Research in Microbiology, 168*(3), 266–275.

Schillberg, S., Raven, N., Fischer, R., Twyman, M., & Schiermeyer, A. (2013). Molecular farming of pharmaceutical proteins using plant suspension cell and tissue cultures. *Current Pharmaceutical Design, 19*(31), 5531–5542.

Schillberg, S., Raven, N., Spiegel, H., Rasche, S., & Buntru, M. (2019). Critical analysis of the commercial potential of plants for the production of recombinant proteins. *Frontiers in Plant Science, 10*, 720.

Spiegel, H., Stöger, E., Twyman, R. M., & Buyel, J. F. (2018). Current status and perspectives of the molecular farming landscape. *Molecular Pharming: Applications, Challenges and Emerging Areas*, 3–23.

Tabashnik, B. E., Gassmann, A. J., Crowder, D. W., & Carrière, Y. (2008). Insect resistance to Bt crops: Evidence versus theory. *Nature Biotechnology, 26*(2), 199–202.

Tschofen, M., Knopp, D., Hood, E., & Stöger, E. (2016). Plant molecular farming: Much more than medicines.

Turnbull, C., Lillemo, M., & Hvoslef-Eide, T. A. (2021). Global regulation of genetically modified crops amid the gene edited crop boom–a review. *Frontiers in Plant Science, 12*, 258.

Twyman, R. M., Stoger, E., & Schillberg, S. (2003). Molecular farming in plants: Host systems and expression technology. *Trends in Biotechnology, 21*(12), 570–578.

Twyman, R. M., Schillberg, S., & Fischer, R. (2005). Transgenic plants in the biopharmaceutical market. *Expert Opinion on Emerging Drugs, 10*, 185–218. https://doi.org/10.1517/14728214.10.1.185

Uckun, F. M., Saund, S., Windlass, H., & Trieu, V. (2021). Repurposing anti-malaria phytomedicine artemisinin as a COVID-19 drug. *Frontiers in Pharmacology, 12*, 407.

Vasilev, N., Smales, C. M., Schillberg, S., Fischer, R., & Schillberg, A. (2016). Developments in the production of mucosal antibodies in plants. *Biotechnology Advances, 34*, 77–87. https://doi.org/10.1016/j.biotechadv.2015.11.002

Walsh, G. (2018). Biopharmaceutical benchmarks (2018). *Nature Biotechnology, 36*, 1136–1145. https://doi.org/10.1038/nbt.4305

Woodard, S. L., Mayor, J. M., Bailey, M. R., Barker, D. K., Love, R. T., Lane, J. R., Delaney, D. E., McComas-Wagner, J. M., Mallubhotla, H. D., Hood, E. E., Dangott, L. J., Tichy, S. E., & Howard, J. A. (2003). Maize (*Zea mays*)-derived bovine trypsin: characterization of the first large-scale, commercial protein product from transgenic plants. *Biotechnology and Applied Biochemistry, 38*, 123–130.

Chapter 2
Tools and Techniques Used in Plant Molecular Farming

Plant molecular farming uses a range of plant systems for the generation of important and effective recombinant proteins. There are several alternatives accessible when it comes to selecting the plant expression system, ranging from the choice of plant and expression vector to the promoter that will be employed. Depending upon the choice of these options, there can be vast differences in terms of yield, storage capacity, harvesting proficiency needed, and posttranslational modifications of proteins. As a result, these alternatives must be chosen while keeping in mind the criteria for producing the specific protein.

1   Expression Types

To date, whole plant expression has been the most common and preferred expression platform for the creation of recombinant proteins. It either involves steady transgenic expression in the nucleus or chloroplasts, or transient transgene expression via viral, agrobacterium, or hybrid vectors. The establishment of the stable transgenic lines for recombinant protein production requires certain decisions to be taken regarding the site of gene integration (nuclear or chloroplast), compartmentalization of target proteins (e.g., vacuole, apoplast, cytosol, endoplasmic reticulum), and target tissue for expression (leaves or seeds). These decisions are governed by several factors such as stability of the expressed protein, desired expression levels, posttranslational modifications needed, and downstream processing charges (Hood et al., 2012, b; Xu et al., 2018). A regulated and high degree of transcription, protein folding, translation, targeting, and stability is required to achieve the appropriate yield of recombinant proteins (Burnett & Burnett, 2020). Regulatory genomic elements like promoters and polyadenylation sites are essential for high transcription.

© The Author(s), under exclusive license to Springer Nature Switzerland AG 2022
K. I. Wani, T. Aftab, *Plant Molecular Farming*, SpringerBriefs in Plant Science,
https://doi.org/10.1007/978-3-031-12794-6_2

1.1 Nuclear Expression

Before using any of the plant transformation methods, a DNA source called transgene cassette is designed. The transgene cassette enables the host plant to produce the desired protein. It contains all the regulatory elements which enable the successful expression of the foreign gene and also help in the selection of genetically modified plants. These elements include a foreign for desired protein expression, a promoter sequence for transcription initiation at the right place, a termination sequence to mark the end of the gene sequence, and a marker gene for identification of plant tissues expressing the desired protein.

Integrating foreign genes into the nuclear genome of plants is the most conventional technique for producing transgenic plants for the synthesis of recombinant proteins. The foreign gene in the expression vector can be transformed into the nuclear genome using *Agrobacterium tumefaciens* or particle bombardment (gene gun/biolistic) methods; signal peptides are then employed for protein storage and production (Shadid & Daniell, 2016). From a pool of stable transgenic lines, the best one is selected and used for recombinant protein production for successive generations (Shanmugaraj et al., 2020).

1.1.1 Agrobacterium-Mediated Transformation

Different transformation methods are chosen depending upon the type of plant to be transformed. The Agrobacterium-mediated transformation, which uses the soil bacterium (*Agrobacterium tumefaciens*) to transmit the foreign gene of interest, is most suitable for dicots like *Nicotiana tabacum* and *Nicotiana benthamiana* (Heidari Japelaghi et al., 2018; Fischer and Buyel, 2020). The potential obstacle in the transformation of monocots by this bacterium is that the monocots do not produce or produce a very less quantity of acetosyringone, a phenolic compound produced by the dicots upon wounding. These substances are required for activating the vir gene on the Ti (Tumor-inducing) plasmid, allowing T-DNA to be transferred to the plant genome (Mirzaee et al., 2022; Peyret & Lomonossoff, 2015; Smith & Hood, 1995). Moreover, the monocots are also difficult to regenerate after the transformation process.

Under natural conditions, *A. tumefasciens* causes crown gall disease upon entering the wounds present in the host tissue (Gelvin, 2003; Mashiguchi et al., 2019). It transfers a loop of DNA called Ti plasmid into the plant. The Ti plasmid contains two parts- T-DNA, which integrates into the host's nuclear genome, and vir genes, which aid in the transfer and incorporation of T-DNA into the host's nuclear genome (Hwang et al., 2017). After the integration, the host plant is said to be transformed (Fig. 2.1). While designing a Ti-plasmid, its tumor-inducing region (T-DNA) is removed and the transgene cassette carrying the foreign gene of interest along with other regulatory elements is inserted between the intervening sections of T-DNA. The agrobacterium is put through a process of genetic alteration so that it only carries the modified Ti-plasmid ready for transfer into the host plant. After this, the agrobacterium containing the altered Ti-plasmid with transgene cassette is incubated with plant fragments for nearly two days.

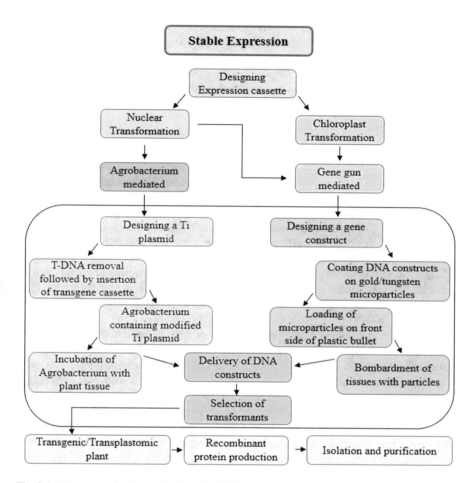

Fig. 2.1 Diagrammatic illustration showing different platforms of stable expression used in plant molecular farming

1.1.2   Biolistic/Particle-Mediated Transformation

The blasting of DNA fragments into cells with a gene gun is termed biolistics (Biological ballistics), also known as particle-mediated transformation. In this method, the DNA constructs are coated on microparticles made of gold or tungsten (Ismagul et al., 2018). These DNA-coated pellets are then put onto the front of a plastic bullet, which is subsequently fired at the plant sample using pressurized helium gas (Lacroix & Citovsky, 2020). High pressure in the helium chamber breaks the rupture disk forcing the bullet down the shaft. At the end of the shaft, the bullet comes to stop but the DNA-coated particles exit and penetrate the biological sample, resulting in the delivery of DNA constructs (Fig. 2.1). After the successful integration of foreign DNA, the repair process begins to correct the damage caused by biolistic events (Liu et al., 2019). This method of transformation is quite useful as it is not bound by biological constraints and can transform almost all types of plants, unlike *Agrobacterium* which mainly transforms dicots. Moreover, it does not have any limitations regarding the shape, sequence, or size of the DNA (Lacroix & Citovsky, 2020). However, gene gun transformation does have some limitations like the requirement of expensive equipment, multiple site insertion, and causing cell death (Rivera et al., 2012).

After the transformation of plant tissue by either *Agrobacterium* or ballistics, the tissues are transferred to culture plates with a suitable selection condition for the selection of transgenic cells. The transgenic plant tissues which survive are transferred to a nutrient medium with appropriate growth conditions and hormones. Due to the totipotent nature of plant cells, a whole new transgenic plant is formed from the transformed tissue. These plants are then allowed to mature and produce seeds. Generally, the first generation of transgenic plants is evaluated to investigate the inheritance of the desired trait to the next progeny (https://www.uni-weimar.de/kunst-und-gestaltung/wiki/images/Transgenic_Crops.pdf; Last retrieved 22 April. 2022). Due to the stable integration of the transgene, recombinant proteins can be generated in subsequent generations. In the initial stages of genetic transformation plants such as *Arabidopsis thaliana* were used to develop stable transformants (Floss et al., 2008).

Despite being the most widely used method of transformation, the nuclear transformation has certain limitations such as low expression of the recombinant protein, transgene escape via gene flow to non-genetically modified counterparts or wild relatives, gene silencing, and time-consuming backcrosses and plant breeding. These limitations could be overcome by CRISPR-Cas9, a novel gene-editing technique for the addition, removal, or modifications of targeted DNA (Burnett & Burnett, 2020; Jaganathan et al., 2018). CRISPR-Cas9 is extremely robust and efficient with no off-targets with 66–100% genome editing efficiency reported in a recent study involving cotton (Wang et al., 2018). The most extensively used promoter in dicots is CaMV 35S from the Caulimovirus (Cauliflower mosaic virus) and it may be enhanced by duplicating its promoter region (Burnett & Burnett, 2020; Ma et al., 2003). Monocots, on the other hand, benefit from promoters like the maize ubiquitin-1 promoter (Ma et al., 2003; Twyman et al., 2003). One of the factors which determine the expression of proteins is polyadenylation which regulates mRNA export from the nucleus to the cytoplasm for translation and is also a vital component of mRNA stability. Several polyadenylation sequences can be used including the pea *ssu* gene, *A. tumefaciens nos* gene, and Caulimovirus (cauliflower mosaic virus) 35S transcript. The protein yield is also governed by the type of promoters used, with strong and constitutive promoters resulting in a high yield of proteins (Ma et al., 2003; Twyman et al., 2003). Furthermore, using tissue-specific promoters like those found in seeds results in the synthesis of target proteins in certain tissues, making harvesting easier and avoiding parent plant toxicity (Twyman et al., 2003). Moreover, using inducible promoters to start recombinant protein synthesis shortly before or after harvest helps to minimize the plant-limiting effects of recombinant protein overexpression (Twyman et al., 2003).

1.2 Chloroplast Expression

The introduction of a transgene into the plastid by particle-mediated transformation is known as chloroplast expression. Chloroplasts have their own set of genetic material called plastome on which the integration of foreign genes takes place. Plastome transformation has certain unique characteristics such as easy and specific gene integration, lack of position effect and gene silencing, high gene expression, polycistronic expression, and prevention of transgene escape due to the maternal inheritance of plastid genome (Ahmad et al., 2016; Jin & Daniell, 2015; Waheed et al., 2015; Zhang et al., 2017). The expression of foreign genes in chloroplasts also allows us to harvest the proteins before reproductive maturity which ensures "total biological containment of transgenes" (Verma et al., 2008). However, it has a narrow host range which mostly includes dicots from the family Solanaceae such as benth and tobacco (Ahmad et al., 2016). Plastid transformation has also been successfully carried out in *Chlamydomonas reinhardtii*, a microalga (Adem et al., 2017). If the chloroplast genome is already sequenced, then its transformation becomes easier via homologous recombination. A transgenic cassette is inserted between two functioning chloroplast genes utilizing two known flanking regions in the chloroplast genome (Daniell et al., 2016).

Due to double-membrane protection around chloroplasts, their transformation is difficult as compared to the transformation of the nuclear genome. However, the use of the gene gun-mediated method has been quite successful in chloroplast transformation (Verma et al., 2008). Furthermore, multiple chloroplasts allow the expression of multiple gene copies in each cell, making protein yield high (around 70% of overall protein content in plant leaves) (Daniell et al., 2016; Shahid & Daniell, 2016).

Agrobacterium-mediated transformation is not an alternative for chloroplasts as the vir proteins of Ti-plasmid inherently direct the T-DNA to the nucleus. Keeping this in mind, other convincing and affordable methods need to be used for chloroplast transformation. One such technique involves the use of polyethylene glycol (PEG) which removes the cell wall, and the cells are then exposed to purified DNA. Although this method is inexpensive as compared to the gene gun mediated method, it has a low transformation efficiency and is difficult to carry out (Waheed et al., 2015; Wani et al., 2010).

After the initial transformation event, heteroplasmic plant cells with some transformed chloroplasts are generated. These cells are subjected to several cycles of antibiotic selection to achieve homoplasmy- a condition in which all the chloroplasts of a cell are transformed. The antibiotic selection causes a drastic reduction in the number of chloroplasts which prompts the rapid division of remaining chloroplasts during cell division (Ahmad et al., 2016). After achieving homoplasmy, the antibiotic marker is excised from the plastome by different methods such as Cre-LoxP recombination (Day & Goldschmidt-Clermont, 2011; Jin & Daniell, 2015). The transplastomic tissue is then cultured to produce mature plants.

The lack of glycosylation in chloroplasts helps in the generation of therapeutic proteins which are glycosylation free and removes their immunogenicity. However, it also hinders the production of therapeutic proteins like antibodies which require glycosylation for their functioning (Verma et al., 2008). Moreover, the lack of glycosylation also provides an opportunity for engineering a custom glycosylation pathway in chloroplasts that does not interact with the host glycosylation which is essential for cell survival. Currently, the chloroplast expression technique is best suited for the generation of recombinant proteins with few posttranslational modifications.

1.3   Transient Expression System

Foreign genes are delivered into plant tissues without stable integration into the genome and then expressed for a limited period, which is known as transient expression (Paul et al., 2013). Transient expression is normally done in dicots as the vectors used are more adapted for use in dicots, with monocot-specific techniques showing less efficiency. It is probably the simplest and most convenient platform for plant molecular farming expression (Rybicki, 2010). The most common transient expression platforms which are currently used are based on plant viruses, *A. tumefasciens*, and hybrid vectors employing the components of both as illustrated in Fig. 2.2.

1.3.1   Agrobacterium-Mediated Transient Expression

In this technique, a Ti plasmid is designed containing a transgene cassette in place of T-DNA. Genetically modified *Agrobacterium* containing this specially designed vector is introduced into leaf cells for infection. Agrobacterium-mediated transient expression work is generally carried out in leaves, but plant roots can also be infected when cultured as cells (Yang et al., 2008). The infection is generally created via vacuum infiltration, syringe agroinfiltration, and wound and agrospray method. In vacuum infiltration, plants in *Agrobacterium* solution are exposed to a vacuum chamber that forces bacteria into leaves. In syringe agroinfiltration, a needleless syringe is used to inject agrobacteria into the abaxial surface of plant leaves where small holes have been incised. In the wound and agrospray method, the leaves are wounded and then sprayed with *Agrobacterium* solution (Chen et al., 2013; Hahn et al., 2015; Komarova et al., 2010). The Ti-plasmid used in the transient expression system is different from the one used in nuclear transformation. It contains alternate forms of some vir genes that prevent T-DNA's integration with the nuclear genome (Peyret & Lomonossoff, 2015). This non-integration of T-DNA ensures its separate existence in the nucleus until it degrades due to environmental or cellular factors.

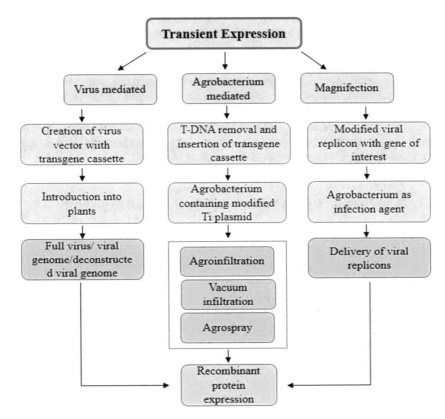

Fig. 2.2 Diagrammatic illustration showing different techniques of transient expression used in plant molecular farming

1.3.2   Virus Mediated Transient Expression

Plant viruses, particularly RNA viruses, are used in this temporary expression method to translate the gene of interest in plant cells (Sainsbury & Lomonossoff, 2008). It relies on the use of mechanical inoculation in which a virus-containing suspension is brought in contact with a wounded leaf (Djikstra & Jagger, 1998). The viral replicating elements are employed to accommodate the gene of interest, which is subsequently episomically amplified and expressed in the cytoplasm of the cell (Lico et al., 2008). In plant molecular farming, the viral genome hijacking property is used to direct the cell to replicate the genome and produce useful proteins. The viral vector gene cassette can be introduced into plants in many forms such as full viruses containing gene cassettes in their genome, viral genomes separated from their protein coats, or deconstructed viral genomes with some viral characters removed (Ibrahim et al., 2019). The viral vectors used for transient expression in plants are usually derived from plant viruses having RNA as their genome. The transgene cassette first needs to be converted from DNA to RNA form for its integration into the RNA genome. The different plant viruses that have been employed contain tobamovirus *(Tobacco mosaic virus)*, Comovirus *(Cowpea mosaic virus),* and potexvirus *(potato virus X(PVX))* (Stoger et al., 2014). Plant viruses have been utilized to produce peptides and polypeptides as fusions to the virus's coat protein, allowing numerous copies of these peptides to be displayed on the surface (Gleba et al., 2005; Lico et al., 2008). The virus-mediated transient expression was also adopted by Large Scale Biology Corporation (Vacaville, USA) which has now been closed, for the generation of patient-specific idiotype vaccines for the therapy of B-cell non-Hodgkin's lymphoma (McCormick et al., 2008).

1.3.3 Magnifection: Hybrid Vector-Mediated Transient Expression

Both agroinfiltration and virus-mediated methods of transient expression have advantages as well as disadvantages. To overcome the drawbacks, a third-generation transfection technology relying on both agrobacteria as well as viruses has been devised. It involves the delivery of viral replicons by utilizing *Agrobacterium* as an infective agent. This eclectic technology is referred to as magnifection and it involves the combination of three biological systems: speed and expression of plant RNA virus, DNA delivery and vector efficiency of *Agrobacterium*, as well as low production cost and posttranslational modifications of plants (Gleba et al., 2005). This hybrid expression system has been used on a commercial scale as well. One of the notable examples is the magnICON® system developed by Icon Genetics which uses a deconstructed *Tobacco mosaic virus* genome and *A. tumefasciens* as a delivery vehicle (Gleba et al., 2005).

1.4 Leaf Based Versus Seed Based Expression

The stable transformation can be leaf-based or seed-based, with tobacco emerging as the most desired plant for leaf-based expression. As it is also a nonfood and nonfeed crop, it has fewer regulatory barriers due to fewer chances of entering the food chain (Twyman et al., 2003). Many other green plants such as clover, alfalfa, and lettuce have also been used as expression platforms. The short shelf life of leafy tissue is one of the drawbacks of leaf-based expression. They need to be processed rapidly to assure product quality and consistency. Some of the limitations of leaf-based expression such as protein storage and stability could be mitigated by using seed-based expression systems (Lau & Sun, 2009). As seeds are naturally suited for protein production and storage with lower protease activity and reduced water content, they form an excellent alternative to the leaf-based protein production systems (Benchabane et al., 2008; Saberianfar et al., 2015). Different seed plants that have been used as expression systems include maize, rice, wheat, barley, soybean, pea, and many others. The edible seeds of rice and maize have GRAS (generally regarded as safe) status, making them excellent candidates for oral vaccine production (Sabalza et al., 2013). Moreover, vaccines and antibodies produced in grain seeds have been discovered to stay viable for years at room temperature (Stoger et al., 2002; Xu et al., 2018). Although seed-based expression platforms have several advantages over leaf-based expression platforms, it still has some limitations. They have lower biomass as compared to leaves and pose the risk of gene escape via pollen or seed dispersal (Xu et al., 2018).

1.5 Plant Cell Suspension Culture

The technique of plant cell suspension culture involves the extraction of tiny groups of cells from plant tissue, followed by their culturing and development into calluses (clumps of disorganized structural tissue) utilizing plant hormones under controlled conditions. After this, the calluses are genetically modified to produce a certain protein, disintegrated in shake flasks, and then transferred into liquid media for cultivation and product collection (Fig. 2.3). Due to their totipotent nature, these cells can express their full genetic potential and produce a range of secondary metabolites. Plant cells can be propagated by using large bioreactors independent of the soil, climate and field management practices. For the manufacturing of biopharmaceuticals, it is an alternative to mammalian cell culture system. The mass of immature plant cells referred to as callus is distributed and propagated in a liquid media to produce cell suspension cultures with the same production capability as complete plants. Since development of human serum albumin in tobacco cells by plant tissue culture, it has been used to generate a wide spectrum of physiologically

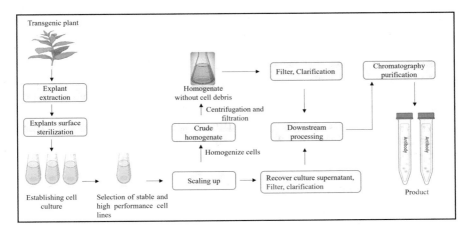

Fig. 2.3 Overview of recombinant protein production in plant cells. For intracellular proteins, homogenization and filtration is necessary to remove coarse debris, whereas for secreted proteins, the product is recovered directly from the culture supernatant

active therapeutic proteins (Sijmons et al., 1990). It has all the advantages of other cell culture-based expression systems but lacks some of the advantages of whole plant-based expression systems such as significant scaling-up potential (Twyman et al., 2003). It guarantees the sterility of *in vitro* conditions and has high-level containment which is suitable for the production of high-purity pharmaceuticals (Franconi et al., 2010). Moreover, the use of low-cost defined growth media as compared to mammalian cell culture and also the cheaper downstream processing and purification make it a viable approach (Kim et al., 2008). The cell suspension-based expression system does not require regeneration and characterization of transgenic plants, which saves a lot of time, making this system rapid as compared to other expression systems used in plant molecular farming (Aviezer et al., 2009). Protalix's recombinant human glucocerebrosidase for the treatment of Gaucher's illness and injectable Retroviral vaccination by Dow Agroscience for chicks are two instances of plant suspension-based pharmaceutical manufacture (Obembe et al., 2011a, b; Shaaltiel et al., 2007). The most extensively used plant cell lines for recombinant protein biosynthesis are derived from tobacco, particularly the BY-2 cultivar. The cells of the BY-2 cultivar easily undergo Agrobacterium-mediated transformation, and cell cycle synchronization and show a robust rate of division (100-fold in a week) (Ozawa & Takaiwa, 2010; Xu et al., 2011, 2018). Just like BY-2 cells, the rice cell suspension cultures are also commonly employed because of a sugar-sensitive α-amylase promoter system (RAmy3D) (Trexler et al., 2005). This promoter is induced by sugar deprivation and results in the increased expression of several pharmacological proteins in rice cells, such as human serum albumin (Liu et al., 2015), interleukin-12 (Shin et al., 2010), hGM-CSF, and α1-antitrypsin (rAAT) (McDonald et al., 2005). However, in contrast to BY-2 cells, the stability of rice cells is less and their vitality is greatly reduced when grown in a sucrose-depleted media (Huang & McDonald, 2009).

Despite its benefits over other expression platforms, it is still not the most popular plant system since the level of recombinant proteins decreases during the stationary phase as a result of increasing proteolytic activity (Corrado & Karali, 2009). Furthermore, it is restricted to a small number of plant cell lines, including rice, carrot, Arabidopsis, alfalfa, and tobacco (Breyer et al., 2009). The major block to the commercialization of plant cell culture technology on a large scale is its low productivity. To avoid these limitations, strategies at process development and molecular levels are essential to enhance the efficiency at all phases of the production pipeline (Xu et al., 2018). Besides productivity challenges, some other challenges regarding plant cell culture include cell culture scale-up in bioprocesses, genetic instability, and non-mammalian glycosylation (James & Lee, 2006; Shih & Doran, 2009).

1.6   Hairy Root Based Expression

Hairy roots are produced by the infection of plant cells with *Rhizobium rhizogenes* (formerly known as *A. rhizogenes*) which contains a root-inducing plasmid termed *Ri Plasmid* (Srivastava & Srivastava, 2007). Just like cell suspension culture, hairy roots may be cultivated under sterilized and regulated conditions for producing pharmaceutical proteins by following cGMP (Current Good Manufacturing Practices). They have some advantages like phenotype and genotype stability as well as plant hormone autotrophy (Georgiev et al., 2012). Infecting wild-type plants with genetically modified *R. rhizogenes* harboring binary vectors encoding the gene of interest or stable transformation of plants (expressing the target protein) with *R. rhizogenes* can produce hairy roots that produce a particular recombinant protein (Pham et al., 2012). Several recombinant proteins have been synthesized by Hairy root cultures after the successful production of a full-length mouse IgG1 in tobacco hairy roots (Wongsamuth & Doran, 1997). Several other recombinant proteins have been produced using hairy root cultures which include growth factors, cytokines, enzymes (e.g., tPA and human acetylcholinesterase), sensor proteins (e.g., GFP and GUS), and antigens (e.g., cholera toxin B surface protective antigen) (Xu et al., 2018). The expression of the human glycogen debranching enzyme involved in the debranching of glycogen has successfully been carried out in *N. benthamiana* plants and hairy root cultures (Rodriguez-Hernandez et al., 2020). It will be beneficial in developing novel biopharmaceuticals based on recombinant GDE protein that might one day be used to treat glycogen storage disorder III. Hairy roots are characterized by rhizosecretion in which the expressed proteins are secreted from the cultured tissues, offering a low-cost and simple purification approach (Guillon et al., 2006; Hood et al., 2012, b). As the purification of recombinant proteins does not require the destruction of root tissue, a given culture can be used for several cycles of production. The major drawback of hairy root technology is low protein productivity (Georgiev et al., 2012). Furthermore, culture scale-up in bioreactors is complicated by the extremely branching phenotypic and non-homogenous growth (Ono & Tian, 2011).

1.7   Moss Based Expression

Moss protonema-based suspension cultures have also been proposed to act as efficient platforms for the production of recombinant proteins. It only requires inorganic salts, light, and water as compared to plant cells which require a sugar-based medium without the light requirement. As it involves differentiated cultures rather than undifferentiated plant cell cultures, it shows genetic stability over long periods (von Stackelberg et al., 2006). The most widely used moss species in molecular farming is *Physcomitrella patens*, whose genome has been completely sequenced in 2006 (http://www.cosmoss.org/). Its sequence can also be easily engineered via gene targeting (Schillberg et al., 2013). Characterization and genome engineering of transgenic strains is eased by the completion of moss genome sequencing. As plants perform posttranslational modifications equivalent to those of mammals, plant-specific *N*-glycosylation of proteins forms a major drawback in plant-based pharmaceutical production due to its immunogenicity. Moreover, transient transfection and efficient secretary signals allow the secretion of recombinant proteins into the surrounding medium (Decker & Reski, 2008).

Glyco-engineering of moss-based proteins through gene targeting involving knocking-in or knocking-out of certain glycosyltransferase has been used for the production of certain humanized glycoproteins (Schuster et al., 2007). Some mosses have been engineered with genes encoding plant-specific glycosyltransferases knocked out of the moss genome (Parsons et al., 2012; Xu et al., 2018), and further, the gene encoding β-1,4-galactosyltransferase is spliced into the fucosyltransferase or xylosyltransferase locus (Huether et al., 2005). Moreover, a gene involved in prolyl hydroxylation has been silenced from the moss genome to prevent any unwanted O-glycosylation at the hydroxyproline residues of human proteins (Parsons et al., 2013).

Some of the recombinant biopharmaceuticals derived from the mosses are of superior quality as compared to those produced via mammalian cell cultures ("bio-betters"). Moss-derived α-galactosidase lacks the terminal mannose phosphate and therefore is taken up by cells via mannose receptors rather than mannose-6 phosphate receptors, yielding better pharmacokinetics in Fabry mice (Reski et al., 2015).

2   Optimization of Protein Expression Levels

The main limitation of plant molecular farming is the low degree of the expression level of recombinant proteins in various expression platforms. Several strategies have been used for the enhanced expression of recombinant proteins, including the use of strong, tissue-specific promoters, intracellular compartment targeting for protein accumulation in stable form, codon optimization of protein sequences, and evaluation of various species and expression techniques (Gomes et al., 2019). The optimization of different components of the expression construct may help in increasing protein production by weakening the epigenetic processes that inhibit gene expression, enhancing transcription, improving mRNA stability, and improving translation (Twyman et al., 2013). Various strong and constitutive promoters used for enhanced expression of recombinant proteins include maize ubiquitin-1 promoter and CaMV 35S promoter in monocots and dicots, respectively. Furthermore, inducible promoters might be employed to avoid the problem of lethality. Tissue- and organ-specific promoters are also advised for directing transgene expression to a specific tissue or organ. Additionally, transcription factors can be employed to increase transgenic expression by boosting promoters (Obembe et al., 2011a, b).

Protein stability can be improved by directing them to compartments that prevent degradation. Purification and extraction steps for protein isolation and type of glycan structures are also affected by protein targeting. An N-terminal peptide signal that is cleaved for protein secretion into the endoplasmic reticulum can route proteins to the secretory pathway. N-terminal transit peptides can be utilized to target chloroplast proteins that don't require posttranslational modifications to function (Kmiec et al., 2014). Moreover, the transformation of chloroplasts with the gene of interest results in an increased accumulation of targeted proteins. Furthermore, the expression of proteins as translational fusion and oleosin proteins, targeting protein expression to oil bodies of seeds can be used to avoid posttranslational modifications of ER (Boothe et al., 2010). Other intracellular compartments, such as protein-sorting vacuoles need further exploration for the accumulation of recombinant proteins (Ou et al., 2014).

3 Downstream Processing: Steps Involved and Challenges

Downstream processing is the recovery and purification of desired products from a biological matrix. In general, the first stages are intended for the production platform, while the later phases are for the product isolation and purification procedures. During the early days of plant biopharming, researchers mainly focused on establishing the potential of genetically modified plants in producing recombinant proteins. The isolation and purification strategies were greatly neglected, due to which downstream processing has become the most expensive part of plant molecular farming, accounting for nearly 80% of total costs (Wilken & Nikolov, 2012). Downstream processing is also an important part of the regulatory procedure for determining biopharmaceutical safety (Fischer et al., 2012). In general, the downstream processing processes follow similar plans and other expression platforms. Tissue harvesting, protein extraction, purification, and formulation are among them (Chen, 2008a, b). In plant biofarming, downstream processing costs are product-specific rather than platform-specific, so the assessment of downstream processing costs and approaches must be done on a case-by-case basis. The biopharmaceutical sector has evolved around a certain number of platforms with similar attributes, and as a result, the early steps of downstream processing have become standardized and include centrifugation and/or purification to eliminate cells as well as other debris. The desired product from the solution is then separated by other processes like filtration and chromatography. The application of this standardized protocol is possible as most biopharmaceutical proteins are secreted by cells into the culture media. It may also be used with plant-based platforms such as cell suspension cultures and entire plant-based systems that secrete proteins into hydroponic conditions. The majority of the plant-derived proteins are retained inside the cells and their extraction requires the disruption of tissues (Wilken & Nikolov, 2012). Moreover, the whole plant-based platforms for biopharmaceutical production must be compatible with excellent manufacturing practices which is a difficult task (Fischer et al., 2012). In some conditions, the products are directed to intracellular organelles for required posttranslational modifications and stability, which makes their extraction compulsory. The external covering of the cell wall also forms a hindrance in the extraction and purification of proteins which are generally secreted into the medium (Buyel et al., 2015). In certain cases, the subcellular accumulation of pharmaceutical proteins might be helpful, particularly when the raw tissues are used for oral administration of prophylactic antibodies or vaccines (Buyel et al., 2015).

Tissue harvesting, protein extraction, purification, and formulation are the primary stages in downstream processing (Chen, 2008a, b). In downstream processing, the initial stages should conform to the production platform, while later steps must be modified to the properties of the protein of interest (Srikanth et al., 2014). The main aim of designing these steps is to maximize the purity and increase the yield of recovered recombinant protein which will in turn save time and reduce expenditure.

3.1 Protein Extraction

This step involves plant/tissue harvesting and the maceration of plant tissues due to which the target protein is released from the biological matrix. In the case of seed-based expression systems, milling is the initial stage. The leaf material or milled seed is then homogenized by using the press or blade-based homogenizers. To make the protein recovery easy, the buffer is added to the biomass which maintains appropriate pH and solubility (Azzoni et al., 2002). To avoid the proteolysis of target proteins, protease inhibitors are also added. Non-protein targets, membrane proteins, and fat removal from biomass are all removed using organic solvents like phenol and hexane. Vacuum infiltration-centrifugation is used for the extraction of extracellular targeted proteins from apoplast and does not require homogenization of whole tissue (Kingsbury & McDonald, 2014). Mechanical extraction procedures can also be replaced by secretion-based technologies. In these systems, the product is secreted into the growth media, from which it is extracted. Rhizosecretion and guttation are some of the other ways of extracting the target product without the destruction of plants.

3.2 Clarification

The ultimate aim of clarification is to remove the contaminants released during tissue homogenization. These impurities include polysaccharides, alkaloids, DNA, RNA, polyphenols, and other proteins released during cell disruption (Woodard et al., 2009). The chromatography apparatus and sensitive filters need extracts of high purity; therefore, the hydroponic medium and culture supernatant also need to be cleared before the purification process. Centrifugation and filtration are frequently used to clarify the extract. Filters are effective and are used only once which minimizes the risk of contamination whereas centrifugation can contaminate the sample. Filter clarification is easier to scale up than centrifugation, making serial filtration the favored approach (Pegel et al., 2011). A densely particulate feed stream can reduce filter capacity which then requires the use of flocculants to improve clarification (Buyel & Fischer, 2014b).

3.3 Flocculation

It's an aggregating process aided by flocculants, which are high-molecular-weight polymers with strong positive or negative charges. They cause aggregation by bridging or neutralization of charges (Buyel & Fischer, 2014a). Dispersed particles form flocks which help in the removal of solid impurities and allow the usage of cheap filters. Flocculation can alter the later steps of downstream processing, so the impact of flocculants needs to be considered before designing the use of flocculants (Buyel & Fischer, 2014b).

3.4 Protein Purification

The major goal of purification is to get a highly purified recombinant protein out of the filtered extract. The different purification techniques are customized for each protein based on its size, charge, solubility, hydrophobicity, pI, or ligand specificity. Purification consists of several sequential chromatography steps. Pretreatment procedures such as precipitation, membrane filtration, and aqueous partitioning can also be employed to increase chromatography performance in the downstream purification phase of downstream processing. Purification should not have more than three processing steps for the most efficient protein rate. However, non-chromatographic technologies, on the other hand, are gaining popularity in large-scale production (Łojewska et al., 2016).

References

Adem, M., Beyene, D., & Feyissa, T. (2017). Recent achievements obtained by chloroplast transformation. *Plant Methods, 13*(1), 1–11.

Ahmad, N., Michoux, F., Lössl, A. G., & Nixon, P. J. (2016). Challenges and perspectives in commercializing plastid transformation technology. *Journal of Experimental Botany, 67*(21), 5945–5960.

Aviezer, D., Brill-Almon, E., Shaaltiel, Y., Hashmueli, S., Bartfeld, D., Mizrachi, S., & Galun, E. (2009). A plant-derived recombinant human glucocerebrosidase enzyme—a preclinical and phase I investigation. *PLoS One, 4*(3), e4792.

Azzoni, A. R., Kusnadi, A. R., Miranda, E. A., & Nikolov, Z. L. (2002). Recombinant aprotinin produced in transgenic corn seed: extraction and purification studies. *Biotechnology and Bioengineering, 80*(3), 268–276.

Benchabane, M., Goulet, C., Rivard, D., Faye, L., Gomord, V., & Michaud, D. (2008). Preventing unintended proteolysis in plant protein biofactories. *Plant Biotechnology Journal, 6*(7), 633–648.

Boothe, J., Nykiforuk, C., Shen, Y., Zaplachinski, S., Szarka, S., Kuhlman, P., Murray, E., Morck, D., & Moloney, M. M. (2010). Seed-based expression systems for plant molecular farming. *Plant Biotechnology Journal, 8*(5), 588–606. https://doi.org/10.1111/j.1467-7652.2010.00511.x

Breyer, D., Goossens, M., Herman, P., & Sneyers, M. (2009). Biosafety considerations associated with molecular farming in genetically modified plants. *Journal of Medicinal Plants Research, 3*(11), 825–838.

Burnett, M. J., & Burnett, A. C. (2020). Therapeutic recombinant protein production in plants: Challenges and opportunities. *Plants People Planet, 2*(2), 121–132.

Buyel, J. F., & Fischer, R. (2014a). Scale-down models to optimize a filter train for the downstream purification of recombinant pharmaceutical proteins produced in tobacco leaves. *Biotechnology Journal, 9*(3), 415–425.

Buyel, J. F., & Fischer, R. (2014b). Flocculation increases the efficacy of depth filtration during the downstream processing of recombinant pharmaceutical proteins produced in tobacco. *Plant Biotechnology Journal, 12*(2), 240–252.

Buyel, J. F., Twyman, R. M., & Fischer, R. (2015). Extraction and downstream processing of plant-derived recombinant proteins. *Biotechnology Advances, 33*(6), 902–913.

Chen, Q. (2008a). Expression and purification of pharmaceutical proteins in plants. *Biological Engineering Transactions, 1*(4), 291–321.

Chen, Q. (2008b). Expression and purification of pharmaceutical proteins in plants. *Biological Engineering Transactions, 1*(4), 291–321.

Chen, Q., Lai, H., Hurtado, J., Stahnke, J., Leuzinger, K., & Dent, M. (2013). Agroinfiltration is an effective and scalable strategy of gene delivery for the production of pharmaceutical proteins. *Advanced Techniques in Biology & Medicine, 1*(1), 103.

Corrado, G., & Karali, M. (2009). Inducible gene expression systems and plant biotechnology. *Biotechnology Advances, 27*(6), 733–743.

Daniell, H., Lin, C. S., Yu, M., & Chang, W. J. (2016). Chloroplast genomes: Diversity, evolution, and applications in genetic engineering. *Genome Biology, 17*(1), 1–29.

Day, A., & Goldschmidt-Clermont, M. (2011). The chloroplast transformation toolbox: Selectable markers and marker removal. *Plant Biotechnology Journal, 9*, 540–553. https://doi.org/10.1111/j.1467-7652.2011.00604.x

Decker, E. L., & Reski, R. (2008). Current achievements in the production of complex biopharmaceuticals with moss bioreactors. *Bioprocess and Biosystems Engineering, 31*(1), 3–9.

Djikstra, J., & Jagger, D. (1998). *Practical plant virology*. Protocol and Exercise.

Fischer, R., Schillberg, S., Hellwig, S., Twyman, R. M., & Drossard, J. (2012). GMP issues for recombinant plant-derived pharmaceutical proteins. *Biotechnology Advances, 30*(2), 434–439.

Fischer, R., & Buyel, J. F. (2020). Molecular farming–the slope of enlightenment. *Biotechnology Advances, 40*, 107519.

Floss, D. M., Sack, M., Stadlmann, J., Rademacher, T., Scheller, J., Stöger, E., & Conrad, U. (2008). Biochemical and functional characterization of anti-HIV antibody–ELP fusion proteins from transgenic plants. *Plant Biotechnology Journal, 6*(4), 379–391.

Franconi, R., Demurtas, O. C., & Massa, S. (2010). Plant-derived vaccines and other therapeutics are produced in contained systems. *Expert Review of Vaccines, 9*(8), 877–892.

Gelvin, S. B. (2003). Agrobacterium-mediated plant transformation: The biology behind the "gene-jockeying" tool. *Microbiology and Molecular Biology Reviews, 67*(1), 16–37.

Georgiev, M. I., Agostini, E., Ludwig-Müller, J., & Xu, J. (2012). Genetically transformed roots: From plant disease to biotechnological resource. *Trends in Biotechnology, 30*(10), 528–537.

Gleba, Y., Klimyuk, V., & Marillonnet, S. (2005). Magnifection—a new platform for expressing recombinant vaccines in plants. *Vaccine, 23*(17–18), 2042–2048.

Gomes, C., Oliveira, F., Vieira, S. I., & Duque, A. S. (2019). Prospects for the production of recombinant therapeutic proteins and peptides in plants: special focus on angiotensin I-converting enzyme inhibitory (ACEI) peptides. In *Genetic engineering-A glimpse of techniques and applications*. IntechOpen.

Guillon, S., Tremouillaux-Guiller, J., Pati, P. K., Rideau, M., & Gantet, P. (2006). Harnessing the potential of hairy roots: Dawn of a new era. *Trends in Biotechnology, 24*(9), 403–409.

Hahn, S., Giritch, A., Bartels, D., Bortesi, L., & Gleba, Y. (2015). A novel and fully scalabla-grobacteriumrium spray-based process for manufacturing cellulases and other cost-sensitive proteins in plants. *Plant Biotechnology Journal, 13*(5), 708–716.

Heidari Japelaghi, R., Haddad, R., Valizadeh, M., Dorani Uliaie, E., & Jalali Javaran, M. (2018). High-efficiency agrobacterium-mediated transformation of tobacco (*Nicotiana tabacum*). *Journal Molecular Plant Breeding, 6*(2), 38–50.

Hood, E., Cramer, C., Medrano, G., & Xu, J. (2012). Protein targeting: Strategic planning for optimizing protein products through plant biotechnology. *Plant Biotechnology and Agriculture,* 35–54. https://doi.org/10.1016/B978-0-12-381466-1.00003-1

Huang, T. K., & McDonald, K. A. (2009). Bioreactor engineering for recombinant protein production in plant cell suspension cultures. *Biochemical Engineering Journal, 45*(3), 168–184.

Huether, C. M., Lienhart, O., Baur, A., Stemmer, C., Gorr, G., Reski, R., & Decker, E. L. (2005). Glyco-engineering of moss lacking plant-specific sugar residues. *Plant Biology, 7*(03), 292–299.

Hwang, H. H., Yu, M., & Lai, E. M. (2017). Agrobacterium-mediated plant transformation: Biology and applications. *Arabic B, 15*, e0186.

Ibrahim, A., Odon, V., & Kormelink, R. (2019). Plant viruses in plant molecular pharming: Toward the use of enveloped viruses. *Frontiers in Plant Science, 10*, 803.

Ismagul, A., Yang, N., Maltseva, E., Iskakova, G., Mazonka, I., Skiba, Y., & Langridge, P. (2018). A biolistic method for high-throughput production of transgenic wheat plants with single gene insertions. *BMC Plant Biology, 18*(1), 1–8.

Jaganathan, D., Ramasamy, K., Sellamuthu, G., Jayabalan, S., & Venkataraman, G. (2018). CRISPR for crop improvement: An updated review. *Frontiers in Plant Science, 9*, 985.

James, E., & Lee, J. M. (2006). Loss and recovery of protein productivity in genetically modified plant cell lines. *Plant Cell Reports, 25*(7), 723–727.

Jin, S., & Daniell, H. (2015). The engineered chloroplast genome just got smarter. *Trends in Plant Science, 20*(10), 622–640.

Kim, T. G., Baek, M. Y., Lee, E. K., Kwon, T. H., & Yang, M. S. (2008). Expression of human growth hormone in transgenic rice cell suspension culture. *Plant Cell Reports, 27*(5), 885–891.

Kingsbury, N. J., & McDonald, K. A. (2014). Quantitative evaluation of E1 endoglucanase recovery from tobacco leaves using the vacuum infiltration-centrifugation method. *BioMed Research International. 2014*, 483596.

Kmiec, B., Teixeira, P. F., & Glaser, E. (2014). Shredding the signal: Targeting peptide degradation in mitochondria and chloroplasts. *Trends in Plant Science, 19*(12), 771–778. https://doi.org/10.1016/j.tplants.09.004. Epub 2014 Oct 7.

Komarova, T. V., Baschieri, S., Donini, M., Marusic, C., Benvenuto, E., & Dorokhov, Y. L. (2010). Transient expression systems for plant-derived biopharmaceuticals. *Expert Review of Vaccines, 9*(8), 859–876.

Lacroix, B., & Citovsky, V. (2020). Biolistic approach for transient gene expression studies in plants. *Biolistic DNA Delivery Plants*, 125–139. https://doi.org/10.1007/978-1-0716-0356-7_6. PMID: 32277451; PMCID: PMC7217558.

Lau, O. S., & Sun, S. S. (2009). Plant seeds as bioreactors for recombinant protein production. *Biotechnology Advances, 27*(6), 1015–1022.

Lico, C., Chen, Q., & Santi, L. (2008). Viral vectors for production of recombinant proteins in plants. *Journal of Cellular Physiology, 216*(2), 366–377.

Liu, Y. K., Li, Y. T., Lu, C. F., & Huang, L. F. (2015). Enhancement of recombinant human serum albumin in transgenic rice cell culture system by cultivation strategy. *New Biotechnology, 32*(3), 328–334.

Liu, J., Nannas, N. J., Fu, F. F., Shi, J., Aspinwall, B., Parrott, W. A., & Dawe, R. K. (2019). Genome-scale sequence disruption following biolistic transformation in rice and maize. *Plant Cell, 31*(2), 368–383.

Łojewska, E., Kowalczyk, T., Olejniczak, S., & Sakowicz, T. (2016). Extraction and purification methods in downstream processing of plant-based recombinant proteins. *Protein Expression and Purification, 120*, 110–117.

Ma, J. K., Drake, P. M., & Christou, P. (2003). The production of recombinant pharmaceutical proteins in plants. *Nature Reviews. Genetics, 4*(10), 794–805. https://doi.org/10.1038/nrg1177

Mashiguchi, K., Hisano, H., Takeda-Kamiya, N., Takebayashi, Y., Ariizumi, T., Gao, Y., & Kasahara, H. (2019). *Agrobacterium tumefaciens* enhances the biosynthesis of two distinct auxins in the formation of crown galls. *Plant & Cell Physiology, 60*(1), 29–37.

McCormick, A. A., Reddy, S., Reinl, S. J., Cameron, T. I., Czerwinkski, D. K., Vojdani, F., & Levy, R. (2008). Plant-produced idiotype vaccines for the treatment of non-Hodgkin's lymphoma: Safety and immunogenicity in phase I clinical study. *Proceedings of the National Academy of Sciences, 105*(29), 10131–10136.

McDonald, K. A., Hong, L. M., Trombly, D. M., Xie, Q., & Jackman, A. P. (2005). Production of human α-1-antitrypsin from transgenic rice cell culture in a membrane bioreactor. *Biotechnology Progress, 21*(3), 728–734.

Mirzaee, M., Osmani, Z., Frébortová, J., & Frébort, I. (2022). Recent advances in molecular farming using monocot plants. *Biotechnology Advances, 58,* 107913.

Obembe, O. O., Popoola, J. O., Leelavathi, S., & Reddy, S. V. (2011a). Advances in plant molecular farming. *Biotechnology Advances, 29*(2), 210–222. https://doi.org/10.1016/j.biotechadv.2010.11.004. Epub 2010 Nov 27.

Obembe, O. O., Popoola, J. O., Leelavathi, S., & Reddy, S. V. (2011b). Advances in plant molecular farming. *Biotechnology Advances, 29*(2), 210–222.

Ono, N. N., & Tian, L. (2011). The multiplicity of hairy root cultures: Prolific possibilities. *Plant Science, 180*(3), 439–446.

Ou, J., Guo, Z., Shi, J., Wang, X., Liu, J., Shi, B., & Yang, D. (2014). Transgenic rice endosperm as a bioreactor for molecular pharming. *Plant Cell Reports, 33*(4), 585–594.

Ozawa, K., & Takaiwa, F. (2010). Highly efficient agrobacterium-mediated transformation of suspension-cultured cell clusters of rice (Oryza sativa L.). *Plant Science, 179*(4), 333–337.

Parsons, J., Altmann, F., Arrenberg, C. K., Koprivova, A., Beike, A. K., Stemmer, C., & Decker, E. L. (2012). Moss-based production of asialo-erythropoietin devoid of Lewis A and other plant-typical carbohydrate determinants. *Plant Biotechnology Journal, 10*(7), 851–861.

Parsons, J., Altmann, F., Graf, M., Stadlmann, J., Reski, R., & Decker, E. L. (2013). A gene r isesponsible for prolyl-hydroxylation of moss-produced recombinant human erythropoietin. *Scientific Reports, 3*(1), 1–8.

Paul, J., Teh, M. Y. H., Twyman, A. M. R., & Ma, K. C. J. (2013). Target product selection-where can molecular pharming make the difference? *Current Pharmaceutical Design, 19*(31), 5478–5485.

Pegel, A., Reiser, S., Steurenthaler, M., & Klein, S. (2011). Evaluating disposable depth filtration platforms for mAb harvest clarification. *Bioprocess International, 9*(9), 52–56.

Peyret, H., & Lomonossoff, G. P. (2015). When plant virology met agrobacterium: The rise of the deconstructed clones. *Plant Biotechnology Journal, 13*(8), 1121–1135.

Reski, R., Parsons, J., & Decker, E. L. (2015). Moss-made pharmaceuticals: From bench to bedside. *Plant Biotechnology Journal, 13*(8), 1191–1198.

Pham, N. B., Schäfer, H., & Wink, M. (2012). Production and secretion of recombinant thaumatin in tobacco hairy root cultures. *Biotechnology Journal, 7*(4), 537–545.

Rivera, A. L., Gómez-Lim, M., Fernández, F., & Loske, A. M. (2012). Physical methods for genetic plant transformation. *Physics of Life Reviews, 9*, 308–345.

Rodriguez-Hernandez, M., Triggiani, D., Ivison, F., Demurtas, O. C., Illiano, E., Marino, C., & Massa, S. (2020). Expression of a functional recombinant human glycogen debranching enzyme (hGDE) in N. benthamiana plants and hairy root cultures. *Protein and Peptide Letters, 27*(2), 145–157.

Rybicki, E. P. (2010). Plant-made vaccines for humans and animals. *Plant Biotechnology Journal, 8*(5), 620–637.

Sabalza, M., Vamvaka, E., Christou, P., & Capell, T. (2013). Seeds as a production system for molecular pharming applications: Status and prospects. *Current Pharmaceutical Design, 19*(31), 5543–5552.

Saberianfar, R., Joensuu, J. J., Conley, A. J., & Menassa, R. (2015). Protein body formation in leaves of *Nicotiana benthamiana*: a concentration-dependent mechanism influenced by the presence of fusion tags. *Plant Biotechnology Journal, 13*(7), 927–937.

Sainsbury, F., & Lomonossoff, G. P. (2008). Extremely high-level and rapid transient protein production in plants without the use of viral replication. *Plant Physiology, 48*(3), 1212–1218.

Schillberg, S., Raven, N., Fischer, R., Twyman, M., & R., & Schiermeyer, A. (2013). Molecular farming of pharmaceutical proteins using plant suspension cell and tissue cultures. *Current Pharmaceutical Design, 19*(31), 5531–5542.

Schuster, M., Jost, W., Mudde, G. C., Wiederkum, S., Schwager, C., Janzek, E., & Gorr, G. (2007). In vivo glyglycoengineeredtibody with improved lytic potential produced by an innovative non-mammalian expression system. *Biotechnology Journal: Healthcare Nutrition Technology, 2*(6), 700–708.

Shaaltiel, Y., Bartfeld, D., Hashmueli, S., Baum, G., Brill-Almon, E., Galili, G., & Aviezer, D. (2007). Production of glucocerebrosidase with terminal mannose glycans for enzyme replacement therapy of Gaucher's disease using a plant cell system. *Plant Biotechnology Journal, 5*(5), 579–590.

Shahid, N., & Daniell, H. (2016). Plant-based oral vaccines against zoonotic and non-zoonotic diseases. *Plant Biotechnology Journal, 14*(11), 2079–2099.

Shanmugaraj, B., Bulaon, I., & C. J., & Phoolcharoen, W. (2020). Plant molecular farming: A viable platform for recombinant biopharmaceutical production. *Planning Theory, 9*(7), 842.

Shih, S. M. H., & Doran, P. M. (2009). Foreign protein production using plant cell and organ cultures: Advantages and limitations. *Biotechnology Advances, 27*(6), 1036–1042.

Shin, Y. J., Lee, N. J., Kim, J., An, X. H., Yang, M. S., & Kwon, T. H. (2010). High-level production of bioactive heterodimeric protein human interleukin-12 in rice. *Enzyme and Microbial Technology, 46*(5), 347–351.

Sijmons, P. C., Dekker, B. M., Schrammeijer, B., Verwoerd, T. C., Van Den Elzen, P. J., & Hoekema, A. (1990). Production of correctly processed human serum albumin in transgenic plants. *Bio/Technology, 8*(3), 217–221.

Smith, R. H., & Hood, E. E. (1995). *Agrobacterium tumefaciens* transformation of monocotyledons. *Crop Science, 35*(2), 301–309.

Srikanth, C. H., Joshi, P., Bikkasani, A. K., Porwal, K., & Gayen, J. R. (2014). Bone distribution study of anti leprotic drug clofazimine in rat bone marrow cells by a sensitive reverse-phase liquid chromatography method. *Journal of Chromatography B, 960*, 82–86.

Srivastava, S., & Srivastava, A. K. (2007). Hairy root culture for mass-production of high-value secondary metabolites. *Critical Reviews in Biotechnology, 27*(1), 29–43.

Stoger, E., Sack, M., Fischer, R., & Christou, P. (2002). Plantibodies: Applications, advantages, and bottlenecks. *Current Opinion in Biotechnology, 13*(2), 161–166.

Stoger, E., Fischer, R., Moloney, M., & Ma, J. K. C. (2014). Plant molecular pharming for the treatment of chronic and infectious diseases. *Annual Review of Plant Biology, 65*, 743–768.

Trexler, M. M., McDonald, K. A., & Jackman, A. P. (2005). A cyclical semicontinuous process for the production of human α1-antitrypsin using metabolically induced plant cell suspension cultures. *Biotechnology Progress, 21*(2), 321–328.

Twyman, R. M., Stoger, E., Schillberg, S., Christou, P., & Fischer, R. (2003). Molecular farming in plants: Host systems and expression technology. *Trends in Biotechnology, 21*(12), 570–578.

Twyman, R. M., Schillberg, S., & Fischer, R. (2013). Optimizing the yield of recombinant pharmaceutical proteins in plants. *Current Pharmaceutical Design, 19*(31), 5486. https://doi.org/1 0.2174/1381612811319310004

Verma, D., Samson, N. P., Koya, V., & Daniell, H. (2008). A protocol for expression of foreign genes in chloroplasts. *Nature Protocols, 3*(4), 739–758. https://doi.org/10.1038/nprot.2007.522

von Stackelberg, M., Rensing, S. A., & Reski, R. (2006). Identification of genic moss SSR markers and a comparative analysis of twenty-four algal and plant gene indices reveal species-specific rather than group-specific characteristics of microsatellites. *BMC Plant Biology, 6*(1), 1–14.

Waheed, M. T., Ismail, H., Gottschamel, J., Mirza, B., & Lössl, A. G. (2015). Plastids: The green frontiers for vaccine production. *Frontiers in Plant Science, 6*, 1005.

Wang, P., Zhang, J., Sun, L., Ma, Y., Xu, J., Liang, S., & Zhang, X. (2018). High efficient multisites genome editing in allotetraploid cotton (*Gossypium hirsutum*) using CRISPR/Cas9 system. *Plant Biotechnology Journal, 16*, 137–150. https://doi.org/10.1111/pbi.12755

Wani, H. S., Haider, N., Kumar, H., & Singh, B. N. (2010). Plant plastid engineering. *Current Genomics, 11*(7), 500–505. https://doi.org/10.2174/138920210793175912

Wilken, L. R., & Nikolov, Z. L. (2012). Recovery and purification of plant-made recombinant proteins. *Biotechnology Advances, 30*(2), 419–433.

Wongsamuth, R., & Doran, P. M. (1997). Production of monoclonal antibodies by tobacco hairy roots. *Biotechnology and Bioengineering, 54*(5), 401–415.

Woodard, S. L., Wilken, L. R., Barros, G. O., White, S. G., & Nikolov, Z. L. (2009). Evaluation of monoclonal antibody and phenolic extraction from transgenic Lemna for purification process development. *Biotechnology and Bioengineering, 104*(3), 562–571.

Xu, J., Ge, X., & Dolan, M. C. (2011). Towards high-yield production of pharmaceutical proteins with plant cell suspension cultures. *Biotechnology Advances, 29*(3), 278–299.

Xu, J., Towler, M., & Weathers, P. J. (2018). Platforms for plant-based protein production. *Bioprocessing of Plant In Vitro Systems*, 509. https://doi.org/10.1007/978-3-319-54600-1_14

Yang, L., Wang, H., Liu, J., Li, L., Fan, Y., Wang, X., & Wang, X. (2008). A simple and effective system for foreign gene expression in plants via root absorption of agrobacterial suspension. *Journal of Biotechnology, 134*(3–4), 320–324.

Zhang, B., Shanmugaraj, B., & Daniell, H. (2017). Expression and functional evaluation of biopharmaceuticals made in plant chloroplasts. *Current Opinion in Chemical Biology, 38*, 17–23.

Chapter 3
Production of Plant Natural Products in Heterologous Microbial Species

1 Introduction

Plants create a variety of compounds, including secondary metabolites with a variety of forms and bioactive capabilities. These products have a long history of being extensively used in the cosmeceutical, pharmaceutical, and food industries, besides being involved in the response of plants to different environmental cues (Maeda, 2019). However, due to the low production of plant natural products in plants, the traditionally used extraction methods and disproportionate exploitation result in the exhaustion of plant resources. Furthermore, most of these plant natural products are structurally complex which makes their chemical synthetic production inefficient and sometimes impossible (Morrison & Hergenrother, 2013). However, the flourishing omics-based strategies and synthetic biology have made the production of these products a great success in engineered hosts.

Besides their role in plant growth and survival, plant metabolites have different medicinal properties. A diverse range of plant-based natural products is taken as part of the human diet, providing a variety of health advantages such as anticancer and nutritional value (Crozier et al., 2008; Seca & Pinto, 2018). Researchers have shown immense interest in the commercialization of these products which can be done either by increasing the production of desirable compounds or reducing the production of unfavorable compounds for the increased flux of energy and substrates towards the desired pathway. Moreover, the use of microorganisms as production "factories" by engineering them to produce the desired compounds has been gaining wide attention in the nutraceutical and pharmaceutical industries (Birchfield & McIntosh, 2020).

Identification of target genes and transfer to hosts other than the source for the synthesis of needed proteins is known as heterologous expression. As protein separation and purification from plant sources can be expensive and time-consuming, a heterologous expression system utilizing microbial species is a good and practical alternative. The ease of genetic manipulation and well-characterized metabolic backgrounds make microorganisms such as *Escherichia coli* desired platforms for the manufacture of plant natural products (Ajikumar et al., 2010). The sustainable production of these products requires a substantial amount of work to engineer these microbes, owing to the strict regulatory mechanisms in different metabolic pathways. The standard strategies for the synthesis of plant natural products in microbial systems include host selection and engineering, design and reconstruction of metabolic processes, management of metabolic engineering, and protein engineering (Birchfield & McIntosh, 2020; Crozier et al., 2008). A well-balanced supply of enzymes, ATP, cofactors, and other metabolites is required for the expression of metabolites in biological systems. Substantial progress achieved in the metabolic reconstitution of microbial systems for the synthesis of various plant natural products (Pyne et al., 2019). Some of the important plant metabolites engineered in microbial hosts have been summarized in Table 3.1.

Table 3.1 Examples of modified microbial strains which produce a variety of medicinally significant chemicals or can be utilized as platform strains for the synthesis of important downstream molecules

Compound type	Example	Bioactivity/ Utility	Native host	Heterologous host	Reference
Monoterpene	Alpha-pinene	Antimicrobial, anti-inflammatory	*Pinus taeda*	*Escherichia coli*	Xu et al. (2014)
Diterpene	Ferruginol Taxadiene	Antibacterial, anticancer Precursor of taxol	*Salvia miltiorrhiza Taxus brevilifolia*	*Saccharomyces cerevisiae Escherichia. Coli*	Guo et al. (2013); Ajikumar et al. (2010)
Sesquiterpene	Alpha-farnesene Amorpha-4,11-diene Amorpha-4,11-Diene Amorpha-4,11-Diene Artemisinin	Insect repellent, pheromone Artemisinin precursor Artemisinin precursor Artemisinin precursor Antimalarial, anticancer, antitumor	*Malus Domestica Artemisia annua Artemisia annua Artemisia annua Artemisia annua*	*Escherichia coli Saccharomyces cerevisiae Escherichia coli Bacillus subtilis Saccharomyces cerevisiae*	Wang et al., (2011a, b); Ro et al. (2006); Wu et al. (2011); Pramastya et al. (2021); Elfahmi et al. (2021)
Glycoside	Protopanaxadiol	Antiviral, anticancer	*Panax ginseng*	*Yarrowia lipolytic*	Wu et al. (2019)
Polyphenol	Resveratrol	Antioxidant, anti-inflammatory, antitumor, antiviral	Grapes, berries, passion fruit, white tea plants	*Escherichia. Coli Corynebacterium. glutamicum*	Lim et al. (2011); Camacho-Zaragoza et al. (2016a, b); Kallscheuer et al. (2016)

The metabolic engineering of microbial hosts for the generation of plant natural products depends on repetitive cycles of design, construction, and testing referred to as the DBT cycle (Nielsen & Keasling, 2016). At the pathway level, DBT includes the determination of the biosynthetic route of plant products, and candidate genes/ enzymes are discovered or tested; at the host level, it entails selecting the host and engineering it to produce the precursors of plant products in a sufficient amount; and at the enzyme level, protein engineering may be required for improved function or production of derivative substances.

2 Selection of Suitable Host

A natural substance from plants that is chosen as a metabolic engineering target should have some industrial, medical, or scientific value. The initial step towards the heterologous generation of a plant product is the identification and selection of the appropriate host species for pathway engineering. The previously developed strains of a species in which the required metabolites are produced in abundance may accelerate the progress. Microbes such as *Saccharomyces cerevisiae* and *E. coli* are the most popular hosts due to the accessibility of a range of established strains, can be maintained easily in culture, and have a wide range of tools for genetic manipulation. Till now most of the plant products produced in microbial systems involve the above-mentioned hosts. Among these two, the first decision a metabolic engineer must make when working with a heterologous host is whether to utilize *S. cerevisiae* or *E.coli* (Cravens et al., 2019). For the production of plant products, *S. cerevisiae* has certain advantages like the presence of cellular microcompartments (e.g., peroxisomes) found in plants and the feasibility of genomic integration due to a high rate of homologous recombination. Furthermore, several plant-based enzymes, such as cytochrome P450s, are transmembrane proteins that need the presence of membranous organelles like the endoplasmic reticulum, which are present in *S. cerevisiae*, for appropriate folding and anchoring. Its importance was demonstrated during a project involving semi-synthetic artemisinin production, in which both *S. cerevisiae* and *E. coli* were tested as potential hosts. Although *E. coli* produced significant titers of amorphadiene, the next stage in the pathway is regulated by $P450_{AMO}$, a plant cytochrome P450 (Tsuruta et al., 2009). High activity of $P450_{AMO}$ could not be produced in *E. coli*, forcing the researchers to turn towards *S. cerevisiae* (Paddon & Keasling, 2014). However, *E. coli* on the other hand has a short doubling time as compared to *S. cerevisiae* and is better adapted to high enzyme expression. Furthermore, it has the availability of a distinct native metabolite pool than *S. cerevisiae*. For example, the availability of a native pathway for several isoprenoid compounds has been utilized to engineer *E.coli* to create taxol precursor taxadiene, and its production was found to be 2400-fold as compared to *S. cerevisiae* engineered for taxadiene production (Ajikumar et al., 2010).

Co-culture involving multiple organisms as hosts for the production of plant natural products offers an alternative to single-host systems. The components of the metabolic pathway are partitioned across hosts of the same or different species in co-culture (Camacho-Zaragoza et al., 2016a, b; Fang et al., 2018). It has some advantages over the single host system like a reduced burden on hosts from the heterologous expression, utilizing the most suited species for a specific enzyme of the pathway. However, it has some limitations such as inefficiency in the diffusion and/or transport of metabolic intermediates between host cells in co-culture and difficulty in balancing the growth of multiple hosts as a single culture (Cravens et al., 2019)

3   Mining Biosynthetic Processes of Plant Natural Products

One of the important steps in constructing a microbial cell factory for the manufacture of plant natural products is to identify the original biosynthetic pathway. The elucidation of these pathways has been made possible due to the development of bioinformatics, molecular biology, and omics technologies. The traditionally used methods for the identification of novel genes involved in the synthesis of plant natural products include RNA interference (RNAi) (Guo et al., 2016), reverse transcription-polymerase chain reaction (RT-PCR), virus-induced gene silencing (VIGS) (Hileman et al., 2005), isotopic tracer methods (Di et al., 2013), and rapid amplification of cDNA ends (RACE) (Xiong et al., 2016). However, these methods are time-consuming, labor-intensive, and consequently costly.

The rapid progress of "omics" analysis, sequencing, and screening technologies has led to the discovery of thousands of genes involved in the generation of plant natural products (Medema & Osbourn, 2016). An example of this is the assembly of *Salvia miltiorrhiza* transcriptome which is believed to provide a valuable resource for the elucidation of the tanshinone biosynthesis process (Xu et al., 2015). Tanshinone is isolated from the roots and rhizomes of *Salvia miltiorrhiza* and is a pharmacologically active constituent used in the treatment of diabetes, arthritis, apoplexy, cardiovascular diseases, and cancer (Fang et al., 2021). Similarly in *Brassica napus,* the genetic network involved in flavonoid biosynthesis has been deciphered by high-density genetic linkage mapping (Qu et al., 2016). Bioinformatics and transcriptomic analysis, combined with the genome sequencing of *Siraitia grosvenorii* has greatly contributed to the elucidation of mogroside V (a noncaloric sweetener) biosynthetic pathway (Itkin et al., 2016). The genes involved in cucurbitacin biosynthesis, which are tetracyclic terpenes generated in members of the Cucurbitaceae family and employed as narcotics, emetics, and antimalarials, have been identified using genome-wide association analysis based on the genomic variation map of several cucumber lines (Shang et al., 2014; Zhou et al., 2016). Therefore, advances in omics and screening technologies have substantially aided in the discovery of genes involved in plant natural product biosynthesis pathways.

The identification of novel genes involved in various biosynthetic pathways in plants has also been facilitated by online databases. For example, Phytozome is a comparison platform for green plant genomics that provides the evolutionary history of plant genes, as well as access to the sequences and functional annotations (Goodstein et al., 2012). The Medicinal Plant Genomics Resource contains transcriptome and metabolome information from medicinal plants. Bioinformatics databases like the Kyoto Encyclopedia of Genes and Genomes (KEGG) have been used to create metabolic pathways of various plant natural products by using knowledge about genes, enzymes, reactions, and their regulation (Ogata et al., 1999). These different databases help in the construction of artificial biosynthetic pathways of plant natural products whose natural route is yet to be elucidated. It is carried out by mixing genes from different species. It is one of the most convenient ways of constructing the biosynthetic pathways of plant natural products without using genetic information from the original plants. Opioid biosynthesis in yeast represents one of the significant accomplishments of synthetic biology of engineering a complex metabolic pathway in microbes. In addition to opioids, a large number of other plant products such as gastrodin (*Gastrodia elata*), salidroside (*Rhodiola Rosea*), raspberry ketone (*Rubus idaeus*), and many more have been synthesized in microbes by recombined artificial pathways (Bai et al., 2014, 2016; Jiang et al., 2018; Wang et al., 2019; Yin et al., 2020).

4 Optimization of Microbial Cells for Production of Plant Natural Products

Even after the elucidation of biosynthetic pathways of plant products, their synthesis in microbial cell factories remains challenging for a variety of reasons. Firstly, there occurs a limited degree of carbon metabolic flow toward their precursors as they are mostly secondary metabolites. It results in a limited supply of precursors for heterologous production of these products. Moreover, as most of these plant natural products possess complex molecular structures, their biosynthesis involves multiple genes which form another bottleneck in their microbial production due to difficulty in catalyzing the activities of these in synthetic systems. Furthermore, the precursor or product accumulation in microbial hosts sometimes results in toxicity or feedback inhibition. For increasing the carbon flux towards their precursors, the metabolic fluxes need to be rewired. To avoid toxicity or feedback inhibition, the target metabolites need to secrete out of the cells by using transport engineering strategies.

4.1 Modulating Metabolic Flux for Adequate Precursor Supply

One of the important classes of plant natural products produced in heterologous microbial systems are terpenoids whose biosynthesis occurs via the mevalonate (MVA) pathway and the 2-C-methyl-d-erythritol 4-phosphate (MEP) pathway.

However, the metabolic flux towards these pathways is restricted in microbial hosts. The over-expression of MVA pathway genes in *S. cerevisiae* using GAL promoter results in an increased metabolic flux (Ro et al., 2006). Moreover, the inhibition of competitive pathways is also one of the strategies for the optimization of the desired pathway. The inhibition of competitive pathways diverts the resources toward the desired pathway. An example of this is the use of an inducible promoter for reducing the expression of the *ERG9* gene involved in ergosterol production, which in turn diverts isopentenyl pyrophosphate toward artemisinic acid production (Westfall et al., 2012).

Another class of plant natural products of high value produced in microbial systems is flavonoids whose major precursors are phenylalanine or tyrosine and malonyl-CoA. In glucose-grown shake-flask cultures, the overexpression of *tyrosine ammonia-lyase* and *chalcone synthase*, as well as knocking out *PDC1*, *PDC5*, *PDC6*, and *ARO3*, as well as the introduction of a tyrosine insensitive 3-deoxy-d-arabinose-heptulosonate-7-phosphate synthase mutant resulted in increased extracellular naringenin (40-fold) (Jiang et al., 2005; Koopman et al., 2012). As a result, metabolic engineering techniques for controlling precursor supply have been shown to improve the output of plant natural products in microbes.

4.2   Enzyme Engineering

The metabolic flux towards plant natural products can also be improved by enzyme engineering for the improvement of their catalytic activity. Enzyme engineering used to increase the catalytic activity of enzymes generally consists of two strategies. The first is based on a logical or semi-rational design. The other is based on directed evolution, which employs techniques such as error-prone PCR (Polymerase Chain reaction), site-specific mutations, and random mutations. The knowledge about the protein structure and catalytic mechanism is the basis of rational design methods. For example, the modification of the N-terminus of CYP725A4 has resulted in the highest titer of oxygenated taxanes in *E. coli* (570 ± 45 mg/L) (Biggs et al., 2016). P450BM3 from *Bacillus megaterium* was engineered using a ROSETTA-based energy minimization approach, allowing for the selective oxidation of amorphadiene and the production of artemisinic-11S,12-epoxide in *E. coli* at titers more than 250 mg/L (Dietrich et al., 2009). The directed evolution approach is one of the most widely utilized methods for changing enzyme catalytic properties. For example, in engineered *S. cerevisiae*, directed evolution of lycopene cyclase retained the phytoene synthase function but inactivated the lycopene cyclase function for improved lycopene production. Moreover, the directed evolution of geranylgeranyl diphosphate synthase resulted in enhanced production of geranylgeranyl pyrophosphate, which increased lycopene production (Xie et al., 2015). Directed evolution or rational engineering strategies may mold a promiscuous enzyme for greater specificity, thermostability, and catalytic activity, to increase the conversion rate of precursors into products.

4.3 Transporter Engineering: Avoiding Toxicity and Feedback Inhibition

As discussed earlier, the excessive buildup of secondary metabolites usually causes toxicity and feedback inhibition, which in turn affects cell growth and biosynthesis of plant products. So, engineering transporters could be one of the ways to avoid this problem (Liu et al., 2017). Transport engineering may aid in the development of microbial cell factories by providing solutions to several problems. The altered expression of the transporters may help in the following ways

1. Avoid autotoxicity and feedback inhibition by exporting the desired products out of the cell.
2. Allow cost-effective purification of final products from the surrounding medium.
3. Make it easier for intermediates to move across intracellular organelles.
4. Import compounds used as substrates by the microbe.
5. Prevent intermediate escape or facilitate re-import of escaped intermediates.

Transport engineering may involve altering the expression of heterologous transporters or endogenous microbial transporters (Belew et al., 2022). Let's take an example of the heterologous production of resveratrol in *S. cerevisiae* and *E. coli*. It is an important medicinal compound with anti-inflammatory, antioxidant, anticancer, and anti-aging properties (Salehi et al., 2018). Its production in heterologous microbial systems is limited by the poor export rate of the final product. The heterologous expression of AraE, a microbial arabinose transporter in *S. cerevisiae* raised resveratrol titer by 2.44-fold (Wang et al., 2011a, b). In *E. coli*, the overexpression of 10 endogenous genes was analyzed, including OmpW and OmpF (outer membrane proteins), MarA (positive expression regulator of AcrAB-TolC efflux pump), and seven other transporters. The overexpression of MarA, OmpW, and OmpF resulted in increased extracellular resveratrol titer, with OmpF showing the highest impact. However, the overexpression of some transporters resulted in a reduction of extracellular resveratrol titer, suggesting the significance of prescreening endogenous transporters (Zhao et al., 2018).

5 Artemisinin and Resveratrol: Heterologous Production in Microbial Hosts

5.1 Artemisinin

Artemisinin is a sesquiterpene lactone molecule characterized by the presence of an endoperoxide bridge and is generated in the glandular secretory trichomes (Fig. 3.1) of *Artemisia annua* plants. It is highly effective against malaria, various tumors and cancers, viruses, and several parasites (Efferth, 2017, 2018; Wani et al., 2021). Artemisinin production in *A. annua* is very low (0.01–1.5% dry weight), making it

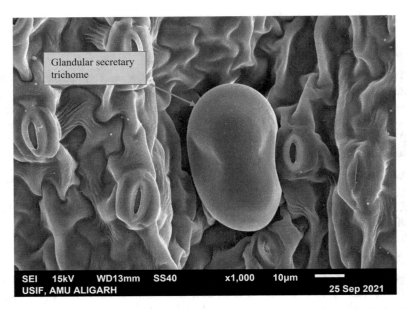

Fig. 3.1 Scanning electron microscopic image of *Artemisia annua* leaf showing glandular secretary trichome

quite costly for the common man and its profitable commercialization a tedious job. The increased artemisinin yield through strategies like breeding and total organic synthesis remains a major challenge. Although the entire synthesis of artemisinin is costly and difficult, the semi-synthesis of this pharmaceutical compound via recombinant microbes offers one of the most promising approaches for bridging the gap between its supply and demand.

Researchers have been making a lot of efforts to enhance the production of artemisinin precursors in microbial systems, with several targets being achieved so far (Ro et al., 2006; Westfall et al., 2012). Engineering of microbial terpene route for artemisinic acid production offers a practical way for the semi-synthesis of artemisinin by biotransformational or chemical methods. Using engineered *S. cerevisiae*, high titers of artemisinin precursor artemisinic acid (100 mg l^{-1}) were produced using an engineered mevalonate process, amorphadiene synthase, and a recombinant cytochrome P450 monooxygenase isolated from *A. annua* (Ro et al., 2006). Following three-step oxidation, artemisinic acid is produced from amorpha-4,11-diene, followed by the export of artemisinic acid into the surrounding media. After recovering artemisinic acid from the surrounding medium, it can be chemically converted into artemisinin (Fig. 3.2) with a yield of 30–40% (w/w) (Roth & Acton, 1989). On a biomass basis, transgenic yeasts may generate equivalent quantities of artemisinic acid as *A. annua* but in a very short time compared to the months-long life cycle of *A. annua*. Furthermore, artemisinic acid's discharge into the extracellular media facilitates its recuperation.

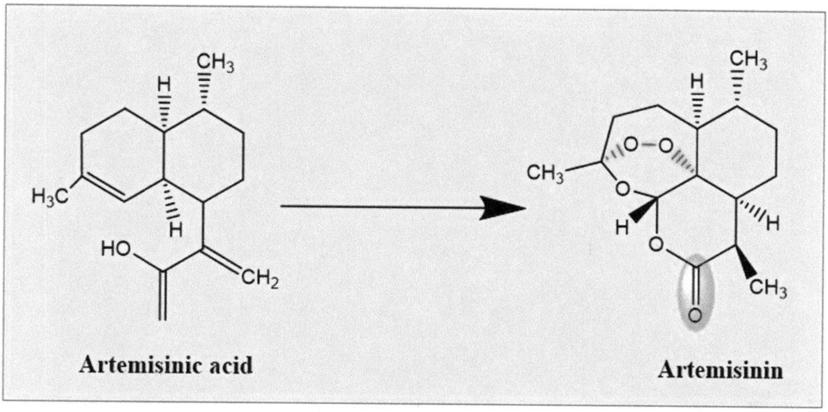

Fig. 3.2 Chemical structures of artemisinic acid and artemisinin. Artemisinin is generally extracted from *Artemisia annua*, but artemisinic acid yield in heterologous microbial systems is higher. Artemisinic acid is then converted into artemisinin

5.2   Resveratrol

Resveratrol (3,5,4′-trihydroxystilbene) is an important plant secondary metabolite found in berries, grapefruit, white tea, and passion fruit. Plants produce it against different microbial infections and stressors. It is an important constituent of different nutritional supplements and pharmaceuticals and possesses a range of biological activities against certain nerve-related diseases and cardiovascular problems (Thapa et al., 2019). Even though resveratrol is generated in plants, its content is generally low, with multistep extraction and seasonal occurrence making it commercially nonviable (Mei et al., 2015). Due to its high pharmacological and dietary value, and to make it commercially viable, its production by alternate methods becomes necessary. Even though it can be produced in large quantities by chemical synthesis, the contamination of resveratrol due to the formation of unwanted side products makes its purification a complicated affair, with an additional risk of using such resveratrol as a food ingredient or medicine (Quideau et al., 2011). Several microbes including both prokaryotes (*E. coli*, *Streptomyces venezuelae*, *Lactococcus lactis*, and *Corynebacterium glutamicum*) and eukaryotes (*S. cerevisiae*) have been used hosts for the heterologous production of resveratrol. Different strategies like metabolic engineering, protein engineering, alternative enzyme selection, system and synthetic biology techniques, and central carbon flow redirection have all facilitated the synthesis of resveratrol in different microbes (Pandey et al., 2016) (Fig. 3.3).

Beekwilder and colleagues used yeast that included stilbene synthase from *Vitis vinifera* and 4-coumarate coenzyme A ligase from *Nicotiana tabacum* to produce resveratrol at a concentration of 6 mg/L (Beekwilder et al., 2006). Engineered *Corynebacterium. glutamicum* (Δ*phdB*, Δ*pcaF*, and Δ*pobA*) was generated to produce resveratrol from *p*-coumaric acid. This expression system resulted in the

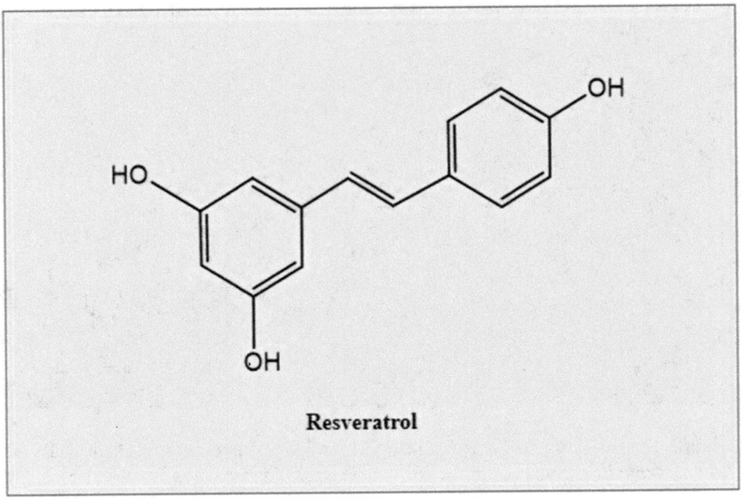

Fig. 3.3 Chemical structure of resveratrol

production of resveratrol at 158 mg/L upon the addition of cerulenin (a fatty acid biosynthesis inhibitor) into the growth medium (Kallscheuer et al., 2016). Several studies have also been carried out for engineering *E. coli* by introducing heterologous genes from plants and other organisms for resveratrol production. *E. coli* engineered with 4-coumarate coenzyme A ligase from *Arabidopsis thaliana* and stilbene synthase from *Arachis hypogaea* resulted in a production of 105 mg/L of resveratrol using 1 mM *p*-coumaric acid as a precursor (Watts et al., 2006). Engineered *E. coli* constructed using a stilbene synthase modification along with an increased supply of intracellular malonyl-CoA resulted in the production of resveratrol titers at a high rate of 2.3 g/L (Lim et al., 2011).

References

Ajikumar, P. K., Xiao, W. H., Tyo, K. E., Wang, Y., Simeon, F., Leonard, E., et al. (2010). Isoprenoid pathway optimization for Taxol precursor overproduction in Escherichia coli. *Science, 330*(6000), 70–74.

Bai, Y., Bi, H., Zhuang, Y., Liu, C., Cai, T., Liu, X., & Ma, Y. (2014). Production of salidroside in metabolically engineered Escherichia coli. *Scientific Reports, 4*(1), 1–8.

Bai, Y., Yin, H., Bi, H., Zhuang, Y., Liu, T., & Ma, Y. (2016). De novo biosynthesis of Gastrodin in *Escherichia coli. Metabolic Engineering, 35*, 138–147.

Beekwilder, J., Wolswinkel, R., Jonker, H., Hall, R., de Vos, C. R., & Bovy, A. (2006). Production of resveratrol in recombinant microorganisms. *Applied and Environmental Microbiology, 72*(8), 5670–5672.

Belew, Z. M., Poborsky, M., Nour-Eldin, H. H., & Halkier, B. A. (2022). Transport engineering in microbial cell factories producing plant-specialized metabolites. *Current Opinion in Green and Sustainable Chemistry, 33*, 100576.

Biggs, B. W., Lim, C. G., Sagliani, K., Shankar, S., Stephanopoulos, G., De Mey, M., & Ajikumar, P. K. (2016). Overcoming heterologous protein interdependency to optimize P450-mediated Taxol precursor synthesis in Escherichia coli. *Proceedings of the National Academy of Sciences, 113*(12), 3209–3214.

Birchfield, A. S., & McIntosh, C. A. (2020). Metabolic engineering and synthetic biology of plant natural products–a minireview. *Current Plant Biology, 24*, 100163.

Camacho-Zaragoza, J. M., Hernández-Chávez, G., Moreno-Avitia, F., Ramírez-Iñiguez, R., Martínez, A., Bolívar, F., & Gosset, G. (2016a). Engineering of a microbial coculture of Escherichia coli strains for the biosynthesis of resveratrol. *Microbial Cell Factories, 15*(1), 1–11.

Camacho-Zaragoza, J. M., Hernández-Chávez, G., Moreno-Avitia, F., Ramírez-Iñiguez, R., Martínez, A., Bolívar, F., & Gosset, G. (2016b). Engineering of a microbial coculture of Escherichia coli strains for the biosynthesis of resveratrol. *Microbial Cell Factories, 15*(1), 1–11.

Cravens, A., Payne, J., & Smolke, C. D. (2019). Synthetic biology strategies for microbial biosynthesis of plant natural products. *Nature Communications, 10*(1), 1–12.

Crozier, A., Clifford, M. N., & Ashihara, H. (2008). *Plant secondary metabolites: Occurrence, structure and role in the human diet.* John Wiley & Sons.

Di, P., Zhang, L., Chen, J., Tan, H., Xiao, Y., Dong, X., & Chen, W. (2013). 13C tracer reveals phenolic acids biosynthesis in hairy root cultures of Salvia miltiorrhiza. *ACS Chemical Biology, 8*(7), 1537–1548.

Dietrich, J. A., Yoshikuni, Y., Fisher, K. J., Woolard, F. X., Ockey, D., McPhee, D. J., & Keasling, J. D. (2009). A novel semi-biosynthetic route for artemisinin production using engineered substrate-promiscuous P450BM3. *ACS Chemical Biology, 4*(4), 261–267.

Efferth, T. (2017). From ancient herb to modern drug: Artemisia annua and artemisinin for cancer therapy. In *Seminars in cancer biology* (Vol. 46, pp. 65–83). Academic Press.

Efferth, T. (2018). Beyond malaria: The inhibition of viruses by artemisinin-type compounds. *Biotechnology Advances, 36*(6), 1730–1737.

Elfahmi, R. A. H., Chrysanthy, T., Synthiarini, V., Masduki, F. F., Setiawan, A., & Muranaka, T. (2021). Expression of two key enzymes of Artemisinin biosynthesis FPS and ADS genes in *Saccharomyces cerevisiae. Advanced Pharmaceutical Bulletin, 11*(1), 181.

Fang, Z., Jones, J. A., Zhou, J., & Koffas, M. A. (2018). Engineering Escherichia coli co-cultures for the production of curcuminoids from glucose. *Biotechnology Journal, 13*(5), 1700576.

Fang, Z. Y., Zhang, M., Liu, J. N., Zhao, X., Zhang, Y. Q., & Fang, L. (2021). Tanshinone IIA: A review of its anti-cancer effects. *Frontiers in Pharmacology, 11*, 611087.

Goodstein, D. M., Shu, S., Howson, R., Neupane, R., Hayes, R. D., Fazo, J., & Rokhsar, D. S. (2012). Phytozome: A comparative platform for green plant genomics. *Nucleic Acids Research, 40*(D1), D1178–D1186.

Guo, J., Zhou, Y. J., Hillwig, M. L., Shen, Y., Yang, L., Wang, Y., & Huang, L. (2013). CYP76AH1 catalyzes the turnover of miltiradiene in tanshinones biosynthesis and enables heterologous production of ferruginol in yeasts. *Proceedings of the National Academy of Sciences, 110*(29), 12108–12113.

Guo, Q., Liu, Q., A Smith, N., Liang, G., Wang, M. B. (2016). RNA silencing in plants: mechanisms, technologies and applications in horticultural crops. *Current Genomics, 17*(6), 476–489.

Hileman, L. C., Drea, S., de Martino, G., Litt, A., Irish, V. F. (2005). Virus-induced gene silencing is an effective tool for assaying gene function in the basal eudicot species *Papaver somniferum* (opium poppy). *The Plant Journal, 44*(2), 334–341.

Itkin, M., Davidovich-Rikanati, R., Cohen, S., Portnoy, V., Doron-Faigenboim, A., Oren, E., & Schaffer, A. (2016). The biosynthetic pathway of the nonsugar, high-intensity sweetener mogroside V from *Siraitia grosvenorii. Proceedings of the National Academy of Sciences, 113*(47), E7619–E7628.

Jiang, H., Wood, K. V., & Morgan, J. A. (2005). Metabolic engineering of the phenylpropanoid pathway in Saccharomyces cerevisiae. *Applied and Environmental Microbiology, 71*(6), 2962–2969.

Jiang, J., Yin, H., Wang, S., Zhuang, Y., Liu, S., Liu, T., & Ma, Y. (2018). Metabolic engineering of Saccharomyces cerevisiae for high-level production of salidroside from glucose. *Journal of Agricultural and Food Chemistry, 66*(17), 4431–4438.

Kallscheuer, N., Vogt, M., Stenzel, A., Gätgens, J., Bott, M., & Marienhagen, J. (2016). Construction of a *Corynebacterium glutamicum* platform strain for the production of stilbenes and (2S)-flavanones. *Metabolic Engineering, 38*, 47–55.

Koopman, F., Beekwilder, J., Crimi, B., van Houwelingen, A., Hall, R. D., Bosch, D., & Daran, J. M. (2012). De novo production of the flavonoid naringenin in engineered Saccharomyces cerevisiae. *Microbial Cell Factories, 11*(1), 1–15.

Lim, C. G., Fowler, Z. L., Hueller, T., Schaffer, S., & Koffas, M. A. (2011). High-yield resveratrol production in engineered *Escherichia coli*. *Applied and Environmental Microbiology, 77*(10), 3451–3460.

Liu, X., Ding, W., & Jiang, H. (2017). Engineering microbial cell factories for the production of plant natural products: from design principles to industrial-scale production. *Microbial Cell Factories, 16*(1), 1–9.

Maeda, H. A. (2019). Harnessing evolutionary diversification of primary metabolism for plant synthetic biology. *The Journal of Biological Chemistry, 294*(45), 16549–16566.

Medema, M. H., & Osbourn, A. (2016). Computational genomic identification and functional reconstitution of plant natural product biosynthetic pathways. *Natural Product Reports, 33*(8), 951–962.

Mei, Y. Z., Liu, R. X., Wang, D. P., Wang, X., & Dai, C. C. (2015). Biocatalysis and biotransformation of resveratrol in microorganisms. *Biotechnology Letters, 37*(1), 9–18.

Morrison, K. C., & Hergenrother, P. J. (2013). Natural products as starting points for the synthesis of complex and diverse compounds. *Natural Product Reports, 31*(1), 6–14.

Nielsen, J., & Keasling, J. D. (2016). Engineering cellular metabolism. *Cell, 164*(6), 1185–1197.

Ogata, H., Goto, S., Sato, K., Fujibuchi, W., Bono, H., & Kanehisa, M. (1999). KEGG: Kyoto encyclopedia of genes and genomes. *Nucleic Acids Research, 27*(1), 29–34.

Paddon, C. J., & Keasling, J. D. (2014). Semi-synthetic artemisinin: A model for the use of synthetic biology in pharmaceutical development. *Nature Reviews. Microbiology, 12*(5), 355–367.

Pandey, R. P., Parajuli, P., Koffas, M. A., & Sohng, J. K. (2016). Microbial production of natural and non-natural flavonoids: Pathway engineering, directed evolution and systems/synthetic biology. *Biotechnology Advances, 34*(5), 634–662.

Pramastya, H., Xue, D., Abdallah, I. I., Setroikromo, R., & Quax, W. J. (2021). High-level production of amorphadiene using *Bacillus subtilis* as an optimized terpenoid cell factory. *New Biotechnology, 60*, 159–167.

Pyne, M. E., Narcross, L., & Martin, V. J. (2019). Engineering plant secondary metabolism in microbial systems. *Plant Physiology, 179*(3), 844–861.

Qu, C., Zhao, H., Fu, F., Zhang, K., Yuan, J., Liu, L., & Li, J. N. (2016). Molecular mapping and QTL for expression profiles of flavonoid genes in Brassica napus. *Frontiers in Plant Science, 7*, 1691.

Quideau, S., Deffieux, D., Douat-Casassus, C., & Pouységu, L. (2011). Plant polyphenols: Chemical properties, biological activities, and synthesis. *Angewandte Chemie, International Edition, 50*(3), 586–621.

Ro, D. K., Paradise, E. M., Ouellet, M., Fisher, K. J., Newman, K. L., Ndungu, J. M., & Keasling, J. D. (2006). Production of the antimalarial drug precursor artemisinic acid in engineered yeast. *Nature, 440*(7086), 940–943.

Roth, R. J., & Acton, N. (1989). A simple conversion of artemisinic acid into artemisinin. *Journal of Natural Products, 52*, 1183–1185.

Salehi, B., Mishra, A. P., Nigam, M., Sener, B., Kilic, M., Sharifi-Rad, M., & Sharifi-Rad, J. (2018). Resveratrol: A double-edged sword in health benefits. *Biomedicine, 6*(3), 91.

Seca, A. M., & Pinto, D. C. (2018). Plant secondary metabolites as anticancer agents: Successes in clinical trials and therapeutic application. *International Journal of Molecular Sciences, 19*(1), 263.

Shang, Y., Ma, Y., Zhou, Y., Zhang, H., Duan, L., Chen, H., & Huang, S. (2014). Biosynthesis, regulation, and domestication of bitterness in cucumber. *Science, 346*(6213), 1084–1088.

Thapa, S. B., Pandey, R. P., Park, Y. I., & Sohng, J. K. (2019). Biotechnological advances in resveratrol production and its chemical diversity. *Molecules, 24*(14), 2571.

Tsuruta, H., Paddon, C. J., Eng, D., Lenihan, J. R., Horning, T., Anthony, L. C., & Newman, J. D. (2009). High-level production of amorpha-4, 11-diene, a precursor of the antimalarial agent artemisinin, in Escherichia coli. *PLoS One, 4*(2), e4489.

Wang, Y., Halls, C., Zhang, J., Matsuno, M., Zhang, Y., & Yu, O. (2011a). Stepwise increase of resveratrol biosynthesis in yeast Saccharomyces cerevisiae by metabolic engineering. *Metabolic Engineering, 13*(5), 455–463.

Wang, C., Yoon, S. H., Jang, H. J., Chung, Y. R., Kim, J. Y., Choi, E. S., & Kim, S. W. (2011b). Metabolic engineering of Escherichia coli for α-farnesene production. *Metabolic Engineering, 13*(6), 648–655.

Wang, C., Zheng, P., & Chen, P. (2019). Construction of synthetic pathways for raspberry ketone production in engineered Escherichia coli. *Applied Microbiology and Biotechnology, 103*(9), 3715–3725.

Wani, K. I., Choudhary, S., Zehra, A., Naeem, M., Weathers, P., & Aftab, T. (2021). Enhancing artemisinin content in and delivery from Artemisia annua: A review of alternative, classical, and transgenic approaches. *Planta, 254*(2), 1–15.

Watts, K. T., Lee, P. C., & Schmidt-Dannert, C. (2006). Biosynthesis of plant-specific stilbene polyketides in metabolically engineered Escherichia coli. *BMC Biotechnology, 6*(1), 1–12.

Westfall, P. J., Pitera, D. J., Lenihan, J. R., Eng, D., Woolard, F. X., Regentin, R., & Paddon, C. J. (2012). Production of amorphadiene in yeast, and its conversion to dihydroartemisinic acid, a precursor to the antimalarial agent artemisinin. *Proceedings of the National Academy of Sciences, 109*(3), E111–E118.

Wu, T., Wu, S., Yin, Q., Dai, H., Li, S., Dong, F., & Fang, H. (2011). Biosynthesis of amorpha-4, 11-diene, a precursor of the antimalarial agent artemisinin, in Escherichia coli through introducing the mevalonate pathway. *Sheng wu Gong Cheng Xue bao= Chinese Journal of Biotechnology, 27*(7), 1040–1048.

Wu, Y., Xu, S., Gao, X., Li, M., Li, D., & Lu, W. (2019). Enhanced protopanaxadiol production from xylose by engineered *Yarrowia lipolytica. Microbial Cell Factories, 18*(1), 1–12.

Xie, W., Lv, X., Ye, L., Zhou, P., & Yu, H. (2015). Construction of lycopene-overproducing Saccharomyces cerevisiae by combining directed evolution and metabolic engineering. *Metabolic Engineering, 30*, 69–78.

Xiong, S., Tian, N., Long, J., Chen, Y., Qin, Y., Feng, J., & Liu, S. (2016). Molecular cloning and characterization of a flavanone 3-hydroxylase gene from Artemisia annua L. *Plant Physiology and Biochemistry, 105*, 29–36.

Xu, Y., Zhou, T., Espinosa-Artiles, P., Tang, Y., Zhan, J., & Molnár, I. (2014). Insights into the biosynthesis of 12-membered resorcylic acid lactones from heterologous production in Saccharomyces cerevisiae. *ACS Chemical Biology, 9*(5), 1119–1127.

Xu, Z., Peters, R. J., Weirather, J., Luo, H., Liao, B., Zhang, X., & Chen, S. (2015). Full-length transcriptome sequences and splice variants were obtained by a combination of sequencing platforms applied to different root tissues of Salvia miltiorrhiza and tanshinone biosynthesis. *The Plant Journal, 82*(6), 951–961.

Yin, H., Hu, T., Zhuang, Y., & Liu, T. (2020). Metabolic engineering of *Saccharomyces cerevisiae* for high-level production of gastrodin from glucose. *Microbial Cell Factories, 19*(1), 1–12.

Zhao, Y., Wu, B. H., Liu, Z. N., Qiao, J., & Zhao, G. R. (2018). Combinatorial optimization of resveratrol production in engineered E. coli. *Journal of Agricultural and Food Chemistry, 66*(51), 13444–13453.

Zhou, Y., Ma, Y., Zeng, J., Duan, L., Xue, X., Wang, H., & Huang, S. (2016). Convergence and divergence of bitterness biosynthesis and regulation in Cucurbitaceae. *Nature plants, 2*(12), 1–8.

Chapter 4
Sustainable Manufacturing of Vaccines, Antibodies, and Other Pharmaceuticals

The recombinant pharmaceutical protein industry is massive and quickly expanding, with the most significant pharmaceutical firms indicating that these medicines are generating more income than small-molecule pharmaceuticals (Goodman, 2009). Generally, these pharmaceutical compounds are manufactured using bacterial or mammalian cell-based systems, posing a risk of contamination with human pathogens and having complex operating procedures. The production facilities set up for this purpose involve substantial financial risk and require huge capital expenditure. Plant molecular farming has the potential to eliminate some of these limitations, but there are still several hurdles before this industry establishes itself commendably in the pharmaceutical sector. The production of biopharmaceuticals in plants has been one of the most sought-after benefits of plant molecular farming. Biopharmaceuticals are drug products made from living organisms that are used for diagnostic or therapeutic reasons, as well as nutritional supplements.

Plant-made pharmaceuticals (PMPs) are the outcome of a biotechnological breakthrough. In plant molecular farming, plants act as biofactories to produce therapeutic recombinant proteins, which could then be used to treat diseases. The potential of plants as biofactories for the synthesis of recombinant pharmaceutical proteins was first stirred by the biosynthesis of a functional murine monoclonal antibody (mAb) in transformed tobacco lines using "COMPLEMENTARY DNAs" derived from a mouse hybridoma messenger RNA (Hiatt et al., 1989). This provided a stimulus for recombinant pharmaceutical production in plants and was followed by the production of other recombinant proteins including vaccines, antibodies, hormones, growth factors, and therapeutic enzymes (Stoger et al., 2014). Today plants are no longer regarded as medicinal in the classic sense; instead, they are meant to operate as bioreactors for the creation of recombinant proteins to combat various ailments. The emerging field of plant molecular farming has made the plants novel sources of pharmaceutical proteins including antibodies, vaccines, blood substitutes, and other therapeutic compounds (Ahmad et al., 2012; Twyman et al., 2003).

K. I. Wani, T. Aftab, *Plant Molecular Farming*, SpringerBriefs in Plant Science, https://doi.org/10.1007/978-3-031-12794-6_4

1   Plant-Based Vaccines

Since Edward Jenner and Louis Pasteur first understood vaccination, vaccines have been developed primarily by producing (1) pathogen proteins in insect and mammalian cells or (2) microbial pathogens, which are subsequently inactivated and/or purified for final formulations (Flemming, 2020). Recently, vaccines have also been produced by using DNA and RNA (Vetter et al., 2018). The use of plants as biofactories for the production of antigens is another viable approach to carry out this task, which has been presented about three decades ago. During this era, laboratory-tested vaccines have reached clinical use. Although plant-based human vaccines have yet to be commercialized, the production of various viral and bacterial subunit vaccines is being attempted in transgenic plants. The only plant-derived vaccination licensed by the US Department of Agriculture is the Newcastle disease vaccine for poultry. However, several candidate vaccines for human and animal use have reached clinical trials. Currently, the clinical trials of Covid-19 and influenza plant-made vaccines have reached phase 3, with results showing a great promise for their commercialization (Monreal-Escalante et al., 2022).

1.1   Routes of Vaccine Administration

Vaccines can be administered via injection or through mucosal administration. Injections can be subcutaneous or intramuscular whereas mucosal administration can be through oral or nasal routes. Vaccines administered via injections elicit robust protective immunity by promoting IgG synthesis in a preferred manner. They are often effective against infections that infect via the respiratory or systemic pathways; however, such antigens must be purified before use. Such vaccines are commonly made in tobacco plants using a temporary expression technique. Mucosal vaccinations stimulate both mucosal and systemic immunity (Lamichhane et al., 2014). Conceptually, oral plant-based vaccines are considered to be ideal due to their simple manufacturing process; non-requirement of medical devices for injection; and preservation of antigen immunogenicity and biological activities throughout the gastrointestinal tract due to their encapsulation in plant cellular organelles. Mucosal vaccines have all been developed in maize, rice, carrot, lettuce, and potato. Upon reaching the small intestines, these antigens are absorbed into microfold (M) cells in the follicle-associated epithelium for the initiation of systemic and mucosal immune responses (Azegami et al., 2014; Holmgren & Czerkinsky, 2005).

1.2   Edible Vaccines

The notion of edible vaccines has emerged as a novel concept developed by biotechnologists. The edible vaccines also called oral vaccines, food vaccines, green vaccines, and subunit vaccines can be particularly effective in poor and developing countries. The concept of edible vaccines is nothing but the conversion of edible food into potential vaccines for the prevention of infectious diseases by triggering an immune response. Their production involves the introduction of target genes into target plants for the production of desired proteins. The edible vaccine production has found application in the prevention of human as well as animal diseases.

The concept of utilizing plants for the production and delivery of edible subunit vaccines was introduced by Dr. Arntzen in 1990. The first successful attempt in the synthesis of a plant-based edible vaccine was the expression of hepatitis B surface antigen in transgenic tobacco plants by Mason et al. (1992). They transformed tobacco plants with *HBsAg* gene encoding hepatitis B surface antigen, and enzyme-linked immunoassays using a monoclonal antibody directed towards human serum-derived HBsAg indicated the presence of HBsAg in extracts of transformed leaves. Moreover, the level of HBsAg in extracts correlated with the abundance of mRNA, suggesting that this gene has no limitation in its expression at the level of transcription or translation. Upon examination, these recombinant antigens were found physically and antigenically similar to recombinant yeast and human serum-derived HBsAg particles, which were already being used as vaccines. The recombinant HBsAg subunit vaccine produced in tobacco plants by Mason and colleagues is mainly composed of the small protein (S) of hepatitis B virus (HBV), which is identical to the yeast-derived vaccine. The low response of S protein vaccination necessitated the development of effective vaccines against the HBV. The inclusion of surface 1 (PreS1) antigen in the vaccine has been found to enhance its efficacy as previously shown by Xu et al. (1994). To show that HBsAg derived from the plants could initiate an immune response (mucosal) through the oral route, potato tubers were used as an expression system Richter et al. (2000). Qian et al. (2008) used rice for the expression of a novel HBV surface antigen fused with preS1 epitopes, and the resultant recombinant SS1 protein showed an immunological response both against S and preS1 epitopes in BALB/c mice.

1.3 Vaccine Production in Different Plants

Several plants such as maize, rice, tobacco, potato, lettuce, soybean, alfalfa, carrot, and tomato are being utilized as hosts for the introduction of the target genes, which is achieved *in vitro* by cell culture, hairy root culture, and protoplast culture. Chloroplast or nuclear transformation is mainly used to obtain transgenic plants. The technology used and the choice of plant species was used to determine the route of vaccine delivery, as high pressure and heat may destroy the antigens whereas some plants are not edible and need processing. The most desired and sought-after crops for the production of subunit vaccines are cereals due to the long stability of vaccines in stored seeds (Hefferon, 2013). Companies such as Ventria Biosciences and SemBioSys Genetics use seed-based expression systems to produce recombinant proteins.

The edible crops engineered for vaccine production have certain advantages as compared to vaccines produced in plants such as tobacco. In tobacco plants, the antigens need to be purified before their testing, resulting in increased production costs. Furthermore, the maintenance of vaccines in lyophilized conditions removes the necessity of cold chain facilities to store and deliver plant materials. This in turn makes plant-based vaccine production cost-efficient as compared to mammalian or fermentation-based vaccines.

Potato has been used to make vaccines against hepatitis B, diphtheria, tetanus, Norwalk virus, enteritis caused by *E. coli* strain, rabbit hemorrhagic virus, and mink enteritis (Concha et al., 2017; Kurup & Thomas, 2020). Tomato has been reported as an effective vehicle for the expression of Norwalk virus capsid protein for oral immunization (Zhang et al., 2006). Tomato has also been used to produce vaccines against *Vibrio cholera*, bubonic plague, and pneumonia (Alvarez et al., 2006; Sharma et al., 2008). Alfalfa has been investigated as a potential source for vaccinations against *Echinococcus granulosus* and the hog pest virus (Kurup & Thomas, 2020). Tobacco has been tested for the manufacture of vaccines against the Newcastle disease virus, Norwalk virus, HBV, coccidiosis, and a variety of other diseases (Hahn et al., 2007; Herbst-Kralovetz et al., 2010; Sathish et al., 2012). Most of the recent studies on the expression of vaccines have used tobacco because of its ability to produce complex structures with high yields. Some of the vaccines produced in plants have been summarized in Table 4.1.

A model of subunit vaccines called virus-like particles (VLPs) is more immunogenic as compared to the three-dimensional formation of individual antigenic proteins. Since the production of the first VLP by Valenzuela et al. (1982), multiple VLPs have been produced using diverse platforms. A recent model of VLPs has focused on the production of chimeric VLPs, which are made up of antigens from various pathogens. They are also able to integrate functional RNA segments. Chimeric VLPs have certain advantages like multivalence, besides the fact that some antigens and epitopes may serve as adjuvants for others as well. However, there are certain problems regarding their folding that need to be solved. Recently, a plant-derived, quadrivalent, VLP influenza vaccine in adults has been described to be safe and well-tolerated (Pillet et al., 2016).

2 Plant-Based Antibodies

Plants have been presented as appealing host platforms for the manufacture of recombinant proteins, such as mAbs. Besides the scalability, effectiveness, and safety of plants for transgenic protein production, plants can also undergo posttranslational modifications of target proteins, including disulfide bond formation, which is required for the folding and assembling of complex multimeric proteins like mAbs (Basaran & Rodríguez-Cerezo, 2008). Furthermore, plants are also able to perform N-glycosylation, which is comparable to that found in mammals (Gomord & Faye, 2004). However, plant-based recombinant proteins contain core α 1,3-fucose

Table 4.1 Different plant-based vaccines along with the antigens expressed and route of administration

Disease/Pathogen	Antigen	Plant	Administration route	Reference
Malaria (*plasmodium falciparum*)	Pfs25 VLP	Tobacco	Intramuscular	Chichester et al. (2018)
Ebola	Glycoprotein (GP1)	Tobacco	Subcutaneous	Phoolcharoen et al. (2011)
Hepatitis B virus (HBV)	HBsAg	Lettuce	Oral	Kapusta et al. (1999)
HBV	HBsAg and preS2 antigen	Potato	Oral	Joung et al. (2004)
HBV	HBsAg	Potato	Oral	Thanavala et al. (2005)
HBV	SS1 protein	Rice	Intraperitoneal	Qian et al. (2008)
HBV	S/preS1 antigen	Lettuce	Oral	Dobrica et al. (2018)
Enterotoxigenic *Escherichia. coli* (ETEC)	LT-B	Potato	Oral	Tacket et al. (2004)
ETEC	LT-B	Maize	Oral	Tacket et al. (2004)
ETEC	Recombinant fusion protein including CFAB*ST, CFAE, and LTB (CCL)	Tobacco	Oral and subcutaneous	Asmani et al., (2022)
ETEC	MucoRice-CTB	Rice	Oral	Takeyama et al., (2015)
Vibrio cholera	CTB	Rice	Oral	Nochi et al. (2009)
V. Cholera and ETEC	MucoRice-CTB	Rice	Oral	Yuki et al., (2021)
Seasonal influenza	HA Quadrivalent	Tobacco	Intramuscular	Ward et al. (2020)
Influenza (H7N9)	HA (H7)	Tobacco	Intramuscular	Pillet et al. (2015)
Norwalk virus (Gastroenteritis)	Not mentioned	Tobacco	Not available	IconGenetics (Last accessed 18 May, 2022)
COVID-19 (SARS-CoV-2)	Spike protein	Tobacco	Intramuscular	Gobeil et al. (2021)

and terminal bisecting $\beta1,2$-xylose residues but no sialic acid or $\beta1,4$-galactose residues (Gomord et al., 2010; Webster and Thomas, 2012). Plant cells also produce extensin- and arabinogalactan-type O-linked glycans, as opposed to mammalian mucin-type O-linked glycans (Gomord et al., 2010).

Several studies carried out to check the safety and allergenicity of plant-specific glycosylation have shown that they don't have any significant adverse impact on humans (Landry et al., 2010; Tusé et al., 2015). Moreover, plant-specific glycosylation may in turn be advantageous for some recombinant pharmaceutical proteins like taliglucerase alfa (having the brand name Elelyso), which does not require in vitro glycan processing for exposure of terminal mannose glycans for its uptake via macrophage mannose receptors, unlike glucocerebrosidase synthesized in CHO cell line (Shaaltiel et al., 2007). The plant-specific N-linked glycans may also facilitate the uptake of mucosal vaccine antigens by antigen-presenting cells by acting as glycoadjuvants (Gomord et al., 2010).

Despite the studies showing the safety of plant-specific glycans, there are still concerns regarding the therapeutic applications of some plant-based biopharmaceutical proteins such as mAbs, due to the plant glycoepitopes potential immunogenicity in humans (Bosch et al., 2013; Gomord et al., 2010). The introduction of glycoengineered plants that enable mammalian-type glycosylation of recombinant proteins might alleviate these safety issues (Bosch et al., 2013; Gomord et al., 2010). It can be achieved by several approaches which include (a) downregulation of endogenous plant glycan-processing enzymes $\alpha 1,3$-fucosyltransferase and $\beta 1,2$-xylosyltransferase (ΔFT/XT mutants) (Cox et al., 2006; Tusé et al., 2015), (b) using C-terminal H/KDEL (His/Lys-Asp-Glu-Leu) retention peptide for targeting transgenic proteins to the endoplasmic reticulum (Schouten et al., 1996), coexpression of human enzymes for synthesis and addition of mammalian sugars (Rouwendal et al., 2007). Plants like *Nicotiana benthamiana* have been engineered for mammalian mucin-type O-linked glycosylation and terminal sialylation on recombinant proteins (Castilho et al., 2012; Yang et al., 2012). Moreover, in addition to the IgG subtype, functionally active, glycoengineered, heteromultimeric IgM antibodies have also been produced in ΔFT/XT *N. benthamiana* plants (Loos et al., 2014). Furthermore, the *in planta* co-expression of deglycosylating enzymes has also been used to eliminate the N-linked glycans from recombinant proteins (Mamedov et al., 2012).

Besides terrestrial plants, unicellular green algae have also been used to produce recombinant proteins, with chloroplast expression more commonly used for the synthesis of mAbs as compared to nuclear expression (Yusibov et al., 2016. The versatility of chloroplasts of microalgae for the synthesis of recombinant proteins like mAbs is due to large chloroplast size, high expression of target proteins (Rasala & Mayfield, 2015), ease of disulfide bond formation, and isomerase (Yusibov et al., 2016) and chaperone (Schroda, 2004) aided protein folding.

A large number of plant-based manufacturing facilities are needed to fulfill the demand for pharmaceutical proteins, which should comply with current good manufacturing practices (cGMP). Some of the facilities which are already operating have been listed in Fig. 4.1.

Despite these manufacturing facilities, new ones must be built and current ones expanded to guarantee that plant-based systems can have an impact on the mAb market and respond to any pandemic or bioterror threat in the future.

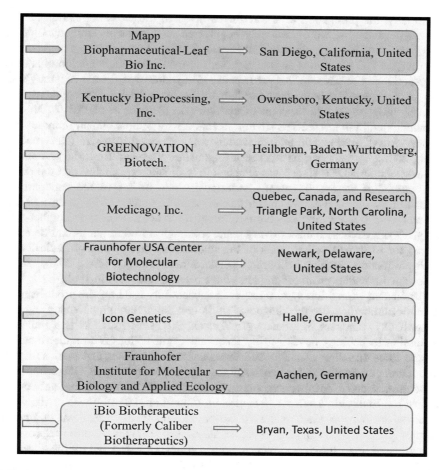

Fig. 4.1 Some of the major plant-based manufacturing facilities for the synthesis of recombinant pharmaceutical proteins

2.1   *Production of mAbs in Plants*

The first study related to the production of a complete antibody (IgG) in plants dates back to nearly 33 years ago when the first plant-based mAb was produced in transgenic tobacco plants by Hiatt and his colleagues in 1989. After this successful attempt, various antibody formats have all been produced in plants which include secretory Fab fragments, IgA, immune cytokines, single-chain antibody fragments, mini bodies, camelid heavy-chain antibodies, and single variable domains. Antibody production in plants has been carried out using several expression platforms based on stable or transient expression of target genes (Sheshukova et al., 2016). The systems used for antibody production in plants include whole plants, cell suspension cultures, and hairy root cultures.

Initially, the production of whole mAbs involved the cross-pollination of two transgenic lines that were separately engineered to express an antibody light or heavy chain. Although it was a laborious task, it enabled the production and correct assembly of IgG molecules as well as other more complex immunoglobulins such as IgAs (Donini & Marusic, 2019; Ma et al., 1995). However, the use of binary vectors carrying genes for both heavy and light chains in the same T-DNA sequence has enabled IgGs' rapid and efficient expression (De Muynck et al., 2010). Human immunoglobulin G mAb 2G12 having potent activity against human immunodeficiency virus-1 *in vitro* and animal models was expressed in maize (Rademacher et al., 2008). To facilitate its storage in organelles produced from the endomembrane system, a KDEL signal was attached to both antibody chains, and the resultant antibodies were subsequently reported in zein bodies of the endoplasmic reticulum. *In vitro* cell assays and surface plasmon resonance analysis showed that the maize-derived antibodies were equivalent to or better than their counterparts derived from Chinese hamster ovary cells. Although different plants can be engineered for mAb production, most of the studies have used tobacco species, with mAb production reaching milligrams per kilogram (Yusibov et al., 2016). Some important antibodies produced in plants have been summarized in Table 4.2.

A transient expression system has been successfully used for the production of chimeric antibodies for the diagnosis of B cell follicular lymphoma using the 'MagnICON' platform in tobacco (Tusé et al., 2015). MagnICON is a transient expression technology developed by Icon Genetics that does not require a stable genetic transformation of plants and thus results in the faster and more efficient production of recombinant proteins. Another application of magnICON technology is the generation of ZMapp, a cocktail of three humanized monoclonal antibodies against the Ebola virus surface glycoprotein (Qiu et al., 2014). These mAbs were produced in tobacco plants and received permission for the treatment of Ebola-infected patients under the FDA's expanded access program. Another expression system based on Comovirus (Cowpea mosaic virus) has been used for the development of deconstructed expression vectors for antibody production. The anti-HIV mAb 2G12 was produced in tobacco plants using this viral-based expression method (Sainsbury et al., 2010). The first mAb to reach clinical trial in Europe was 2G12 produced in tobacco against HIV infection (Ma et al., 2015).

Plant cell suspension cultures have been used to produce recombinant pharmaceutical proteins, particularly antibodies. M12, a tumor-targeting mAb has been obtained from BY-2 cells in an orbitally-shaken disposable bioreactor (Raven et al., 2015). The cell suspension cultures can optimize the glycosylation profile of proteins through glycan metabolic engineering. BY-2 cell lines of tobacco have been engineered to generate antibodies with a glycosylation profile comparable to that of humans (Mercx et al., 2017; Navarre et al., 2017). Hairy roots have also been utilized to make recombinant medicinal proteins in bioreactors. The first reported recombinant pharmaceutical protein produced in hairy roots is Guy's 13, a mAb that targets a vital cellular surface glycoprotein of *Streptococcus mutans* (Wongsamuth & Doran, 1997). Hairy roots have been used to express tumor-targeting mAb H10 either by infecting a glycoengineered *Nicotiana benthamiana* line (ΔXTFT) with

Table 4.2 Different plant-based antibodies and the strategies used for their expression

Host	Clone or product name	Expression strategy	Potential Activity	Reference
N. benthamiana	KPF1	Transgenic	Neutralizing activity against H1N1 influenza A viruses	Park et al. (2020)
N. benthamiana	c2A10G6	Transient	Neutralizes Zika virus	Diamos et al. (2020)
	HSV8	Transient	It traps herpes simplex Virus (HSV) in human cervicovaginal Mucus and neutralizes HSV in mice	Diamos et al. (2020)
N. benthamiana	6D8	Transient	Binds to inactivated Ebola virus	Huang et al. (2010)
N. benthamiana	H10	Transgenic hairy roots	Tumor-targeting	Lonoce et al. (2016)
N. benthamiana	4H2 IgG or IgM	Transgenic hairy roots	Binds chitinase 1antigen	Jugler et al. (2022)
Oryza sativa	2G12	Transgenic	HIV neutralizing antibody	Vamvaka et al. (2016)
Oryza sativa	Cyclic citrullinated peptide	Transgenic	Detection and diagnosis of rheumatoid arthritis	Van Giap et al. (2019)
Oryza sativa	FimA specific mAb	Transgenic	Immunization against *Porphyromonas gingivalis* induced periodontal disease	Kim et al. (2014)
N. Tanacum	scFvG8	Transient	Confers resistance to plants against *Xanthomonas citri*	Raeisi et al. (2022)
N. benthamiana	N6	Transgenic	HIV neutralizing antibody	Moore et al. (2021)

Agrobacterium rhizogenes containing H10 heavy and light chain cDNAs or a transgenic *N. tabacum* line expressing H10 with *A. rhizogenes* (Lonoce et al., 2016). In a recent study, an anti-fungal antibody was produced in Solanaceae hairy roots against *Candida albicans* (Catellani et al., 2020). This antibody was expressed in a single chain (scFvFc) format in hairy roots of *Solanum Lycopersicum* and *N. benthamiana*, followed by testing of culture medium containing the extract against pathogenic fungi. The results showed significant activity of scFvFc 2G8 antibody produced by *N. benthamiana* against *Candida albicans*.

3 Replacement Human Proteins

Replacement human proteins constitute the third important category of plant-based pharmaceutical proteins. These pharmaceutical proteins can be divided into two classes based on their demand in the market. The recombinant pharmaceutical proteins in high demand include blood products such as human serum albumin and replacement proteins such as insulin and gastric lipase. Those with low demand include growth factors/cytokines and recombinant pharmaceuticals for the treatment of rare and orphan diseases such as glucocerebrosidase for Gaucher's disease. Plant-based recombinant pharmaceutical proteins of high demand are appropriate for molecular farming in transgenic plants. Examples of some recombinant pharmaceutical proteins that have achieved clinical development include insulin which is produced by SemBioSys Genetics Inc. in *Carthamus tinctorius* whereas lactoferrin and gastric lipase by Meristem Therapeutics SA are produced in *Zea mays* (Spiegel et al., 2018). Recombinant human precursor insulin was produced in transgenic *Arabidopsis thaliana* seeds and then enzymatically processed *in vitro* to make the final product. The resultant $DesB_{30}$-insulin product exhibited biological activity (Nykiforuk et al., 2006). The pharmaceutical products in low demand have been produced in both plants as well as in mammalian systems. The plant-based glycan structures are another aspect of plant molecular farming that has been exploited for the synthesis of recombinant pharmaceuticals. An example of the exploitation of the plant glycan structures is the production of Locteron, a biobetter form of interferon α2a. Locteron has higher effectiveness, due to the presence of plant glycan structures. Elelyso (Taliglucerase alfa), carrot-derived glucocerebrosidase, benefits from the lack of sialic acid residues on the glycans. The lack of sialic acid residues aids direct absorption of this glucocerebrosidase by macrophages, which are harmed in type 1 Gaucher disease. Contrary to this, glucocerebrosidase generated in Chinese Hamster Ovary cells (Imiglucerase) needs to be trimmed to remove sialic acid residues, which in turn results in its increased production cost. Biologically potent interferon α2b with increased serum half-life and high production has been produced in tobacco cells as arabinogalactan-protein chimeras (Xu et al., 2007). The biological activity of arabinogalactan-protein-interferon α2 chimeras remained similar to that of native interferon 2 with a 13-fold increase in *vivo* serum half-life.

References

Ahmad, P., Ashraf, M., Younis, M., Hu, X., & Kumar, A. (2012). Role of transgenic plants in agriculture and biopharming. *Biotechnology Advances, 30*, 524–540.

Alvarez, M. L., Pinyerd, H. L., Crisantes, J. D., Rigano, M. M., Pinkhasov, J., Walmsley, A. M., & Cardineau, G. A. (2006). Plant-made subunit vaccine against pneumonic and bubonic plague is orally immunogenic in mice. *Vaccine, 24*(14), 2477–2490.

Asmani, F., Khavari-Nejad, R. A., Salmanian, A. H., Amani, J. (2022). Immunological evaluation of recombinant chimeric construct from Enterotoxigenic *E. coli* expressed in hairy roots. *Molecular Immunology, 147*, 81–89.

Azegami, T., Yuki, Y., & Kiyono, H. (2014). Challenges in mucosal vaccines for the control of infectious diseases. *International Immunology, 26*(9), 517–528.

Basaran, P., & Rodríguez-Cerezo, E. (2008). Plant molecular farming: Opportunities and challenges. *Critical Reviews in Biotechnology, 28*(3), 153–172.

Bosch, D., Castilho, A., Loos, A., Schots, A., & Steinkellner, H. (2013). N-glycosylation of plant-produced recombinant proteins. *Current Pharmaceutical Design, 19*(31), 5503–5512.

Castilho, A., Neumann, L., Daskalova, S., Mason, H. S., Steinkellner, H., Altmann, F., & Strasser, R. (2012). Engineering of sialylated mucin-type O-glycosylation in plants. *The Journal of Biological Chemistry, 287*(43), 36518–36526.

Catellani, M., Lico, C., Cerasi, M., Massa, S., Bromuro, C., Torosantucci, A., & Capodicasa, C. (2020). Optimized production of an anti-fungal antibody in Solanaceae hairy roots to develop new formulations against *Candida albicans. BMC Biotechnology, 20*(1), 1–13.

Chichester, J. A., Green, B. J., Jones, R. M., Shoji, Y., Miura, K., Long, C. A., & Yusibov, V. (2018). Safety and immunogenicity of a plant-produced Pfs25 virus-like particle as a transmission-blocking vaccine against malaria: A phase 1 dose-escalation study in healthy adults. *Vaccine, 36*(39), 5865–5871.

Concha, C., Cañas, R., Macuer, J., Torres, M. J., Herrada, A. A., Jamett, F., & Ibáñez, C. (2017). Disease prevention: An opportunity to expand edible plant-based vaccines? *Vaccine, 5*(2), 14.

Cox, K. M., Sterling, J. D., Regan, J. T., Gasdaska, J. R., Frantz, K. K., Peele, C. G., & Dickey, L. F. (2006). Glycan optimization of a human monoclonal antibody in the aquatic plant Lemna minor. *Nature Biotechnology, 24*(12), 1591–1597.

De Muynck, B., Navarre, C., & Boutry, M. (2010). Production of antibodies in plants: Status after twenty years. *Plant Biotechnology Journal, 8*(5), 529–563.

Diamos, A. G., Hunter, J. G., Pardhe, M. D., Rosenthal, S. H., Sun, H., Foster, B. C., & Mason, H. S. (2020). High-level production of monoclonal antibodies using an optimized plant expression system. *Frontiers in Bioengineering and Biotechnology, 7*, 472.

Dobrica, M. O?., Lazar, C., Paruch, L., van Eerde, A., Clarke, J. L., Tucureanu, C., & Branza-Nichita, N. (2018). Oral administration of a chimeric hepatitis B virus S/preS1 antigen produced in lettuce triggers infection-neutralizing antibodies in mice. *Vaccine, 36*(38), 5789–5795.

Donini, M., & Marusic, C. (2019). The current state-of-the-art plant-based antibody production systems. *Biotechnology Letters, 41*(3), 335–346.

Flemming, A. (2020). The origins of vaccination. *Nature Milestones in Vaccines.* S5. Available online: https://media.nature.com/original/magazine-assets/d42859-020-00006-7/d42859-020-00006-7.pdf. (Accessed on 16 May 2022)

Gobeil, P., Pillet, S., Séguin, A., Boulay, I., Mahmood, A., Vinh, D. C., & Landry, N. (2021). Interim report of phase 2 randomized trial of a plant-produced virus-like particle vaccine for Covid-19 in healthy adults aged 18–64 and older adults aged 65 and older. *Medrxiv,* 21257248. https://doi.org/10.1101/2021.05.14.21257248

Goodman, M. (2009). Pharmaceutical industry financial performance. *Nature Reviews Drug Discovery, 8*(12), 927.

Gomord, V., & Faye, L. (2004). Posttranslational modification of therapeutic proteins in plants. *Current Opinion in Plant Biology, 7*, 171–181.

Gomord, V., Fitchette, A. C., Menu-Bouaouiche, L., Saint-Jore-Dupas, C., Plasson, C., Michaud, D., & Faye, L. (2010). Plant-specific glycosylation patterns in the context of therapeutic protein production. *Plant Biotechnology Journal, 8*(5), 564–587.

Hahn, B. S., Jeon, I. S., Jung, Y. J., Kim, J. B., Park, J. S., Ha, S. H., & Kim, Y. H. (2007). Expression of hemagglutinin-neuraminidase protein of Newcastle disease virus in transgenic tobacco. *Plant Biotechnology Reports, 1*(2), 85–92.

Hefferon, K. (2013). Plant-derived pharmaceuticals for the developing world. *Biotechnology Journal, 8*(10), 1193–1202.

Herbst-Kralovetz, M., Mason, H. S., & Chen, Q. (2010). Norwalk virus-like particles as vaccines. *Expert Review of Vaccines, 9*(3), 299–307.

Hiatt, A., Caffferkey, R., & Bowdish, K. (1989). Production of antibodies in transgenic plants. *Nature, 342*(6245), 76–78.

Holmgren, J., & Czerkinsky, C. (2005). Mucosal immunity and vaccines. *Nature Medicine, 11,* S45–S53.

Huang, Z., Phoolcharoen, W., Lai, H., Piensook, K., Cardineau, G., Zeitlin, L., & Chen, Q. (2010). High-level rapid production of full-size monoclonal antibodies in plants by a single-vector DNA replicon system. *Biotechnology and Bioengineering, 106*(1), 9–17.

IconGenetics. Icon genetics clinical development of its novel norovirus vaccine reaches milestone of complete dosing of the first cohort. Available online: https://www.icongenetics.com/icon-genetics-clinical-development-of-its-novel-norovirus-vaccine-reaches-milestone-of-complete-dosing-of-the-first-cohort. (Last accessed May18, 2022).

Joung, Y. H., Youm, J. W., Jeon, J. H., Lee, B. C., Ryu, C. J., Hong, H. J., & Kim, H. S. (2004). Expression of the hepatitis B surface S and preS2 antigens in tubers of Solanum tuberosum. *Plant Cell Reports, 22*(12), 925–930.

Jugler, C., Grill, F. J., Lake, D. F., & Chen, Q. (2022). Humanization and expression of IgG and IgM antibodies in plants as potential diagnostic reagents for valley fever. *bioRxiv*. https://doi.org/10.1101/2022.04.27.489777

Kapusta, J., Modelska, A., Figlerowicz, M., Pniewski, T., Letellier, M., Lisowa, O., & Legocki, A. B. (1999). A plant-derived edible vaccine against hepatitis B virus. *The FASEB Journal, 13*(13), 1796–1799.

Kim, B. G., Kim, S. H., Kim, N. S., Huy, N. X., Choi, Y. S., Lee, J. Y., & Kim, T. G. (2014). Production of monoclonal antibody against FimA protein from Porphyromonas gingivalis in rice cell suspension culture. *Plant Cell Tissue Organ Culture (PCTOC), 118*(2), 293–304.

Kurup, V. M., & Thomas, J. (2020). Edible vaccines: Promises and challenges. *Molecular Biotechnology, 62*(2), 79–90.

Lamichhane, A., Azegami, T., & Kiyono, H. (2014). The mucosal immune system for vaccine development. *Vaccine, 32*(49), 6711–6723.

Landry, N., Ward, B. J., Trépanier, S., Montomoli, E., Dargis, M., Lapini, G., & Vézina, L. P. (2010). Preclinical and clinical development of plant-made virus-like particle vaccine against avian H5N1 influenza. *PLoS One, 5*(12), e15559.

Lonoce, C., Salem, R., Marusic, C., Jutras, P. V., Scaloni, A., Salzano, A. M., & Donini, M. (2016). Production of a tumor-targeting antibody with a human-compatible glycosylation profile in *N. benthamiana* hairy root cultures. *Biotechnology Journal, 11*(9), 1209–1220.

Loos, A., Gruber, C., Altmann, F., Mehofer, U., Hensel, F., Grandits, M., & Steinkellner, H. (2014). Expression and glycoengineering of functionally active heteromultimeric IgM in plants. *Proceedings of the National Academy of Sciences, 111*(17), 6263–6268.

Ma, J. K. C., Hiatt, A., Hein, M., Vine, N. D., Wang, F., Stabila, P., & Lehner, T. (1995). Generation and assembly of secretory antibodies in plants. *Science, 268*(5211), 716–719.

Ma, J. K. C., Drossard, J., Lewis, D., Altmann, F., Boyle, J., Christou, P., & Fischer, R. (2015). Regulatory approval and a first-in-human phase I clinical trial of a monoclonal antibody produced in transgenic tobacco plants. *Plant Biotechnology Journal, 13*(8), 1106–1120.

Mamedov, T., Ghosh, A., Jones, R. M., Mett, V., Farrance, C. E., Musiychuk, K., & Yusibov, V. (2012). Production of non-glycosylated recombinant proteins in Nicotiana benthamiana plants by co-expressing bacterial PNGase F. *Plant Biotechnology Journal, 10*(7), 773–782.

Mason, H. S., Lam, D. M., & Arntzen, C. J. (1992). Expression of hepatitis B surface antigen in transgenic plants. *Proceedings of the National Academy of Sciences, 89*(24), 11745–11749.

Mercx, S., Smargiasso, N., Chaumont, F., De Pauw, E., Boutry, M., & Navarre, C. (2017). Inactivation of the β (1, 2)-xylosyltransferase and the α (1, 3)-fucosyltransferase genes in Nicotiana tabacum BY-2 cells by a multiplex CRISPR/Cas9 strategy results in glycoproteins without plant-specific glycans. *Frontiers in Plant Science, 8*, 403.

Monreal-Escalante, E., Ramos-Vega, A., Angulo, C., & Bañuelos-Hernández, B. (2022). Plant-based vaccines: Antigen design, diversity, and strategies for high level production. *Vaccine, 10*(1), 100.

Moore, C. M., Grandits, M., Grünwald-Gruber, C., Altmann, F., Kotouckova, M., Teh, A. Y. H., & Ma, J. K. C. (2021). Characterization of a highly potent and near pan-neutralizing anti-HIV monoclonal antibody expressed in tobacco plants. *Retrovirology, 18*(1), 1–11.

Navarre, C., Smargiasso, N., Duvivier, L., Nader, J., Far, J., De Pauw, E., & Boutry, M. (2017). N-glycosylation of an IgG antibody secreted by *Nicotiana tabacum* BY-2 cells can be modulated through co-expression of human β-1, 4-galactosyltransferase. *Transgenic Research, 26*(3), 375–384.

Nochi, T., Yuki, Y., Katakai, Y., Shibata, H., Tokuhara, D., Mejima, M., & Kiyono, H. (2009). A rice-based oral cholera vaccine induces macaque-specific systemic neutralizing antibodies but does not influence pre-existing intestinal immunity. *Journal of Immunology, 183*(10), 6538–6544.

Nykiforuk, C. L., Boothe, J. G., Murray, E. W., Keon, R. G., Goren, H. J., Markley, N. A., & Moloney, M. M. (2006). Transgenic expression and recovery of biologically active recombinant human insulin from Arabidopsis thaliana seeds. *Plant Biotechnology Journal, 4*(1), 77–85.

Park, J. G., Ye, C., Piepenbrink, M. S., Nogales, A., Wang, H., Shuen, M., & Kobie, J. J. (2020). A broad and potent H1-specific human monoclonal antibody produced in plants prevents influenza virus infection and transmission in Guinea pigs. *Viruses, 12*(2), 167.

Phoolcharoen, W., Bhoo, S. H., Lai, H., Ma, J., Arntzen, C. J., Chen, Q., & Mason, H. S. (2011). Expression of an immunogenic Ebola immune complex in *Nicotiana benthamiana. Plant Biotechnology Journal, 9*(7), 807–816.

Pillet, S., Racine, T., Nfon, C., Di Lenardo, T. Z., Babiuk, S., Ward, B. J., & Landry, N. (2015). Plant-derived H7 VLP vaccine elicits a protective immune response against the H7N9 influenza virus in mice and ferrets. *Vaccine, 33*(46), 6282–6289.

Pillet, S., Aubin, É., Trépanier, S., Bussière, D., Dargis, M., Poulin, J. F., & Landry, N. (2016). A plant-derived quadrivalent virus-like particle influenza vaccine induces cross-reactive antibody and T cell response in healthy adults. *Clinical Immunology, 168*, 72–87.

Qian, B., Shen, H., Liang, W., Guo, X., Zhang, C., Wang, Y., & Zhang, D. (2008). Immunogenicity of recombinant hepatitis B virus surface antigen fused with preS1 epitopes expressed in rice seeds. *Transgenic Research, 17*(4), 621–631.

Qiu, X., Wong, G., Audet, J., Bello, A., Fernando, L., Alimonti, J. B., & Kobinger, G. P. (2014). Reversion of advanced Ebola virus disease in nonhuman primates with ZMapp. *Nature, 514*(7520), 47–53.

Rademacher, T., Sack, M., Arcalis, E., Stadlmann, J., Balzer, S., Altmann, F., Stoger, E. (2008). Recombinant antibody 2G12 produced in maize endosperm efficiently neutralizes HIV-1 and contains predominantly single-GlcNAc N-glycans. *Plant Biotechnology Journal, 6*(2), 189–201.

Raeisi, H., Safarnejad, M. R., Alavi, S. M., Farrokhi, N., & Elahinia, S. A. (2022). Transient expression of an scFvG8 antibody in plants and characterization of its effects on the virulence factor pthA of Xanthomonas citri subsp. citri. *Transgenic Research, 31*(2), 269–283.

Rasala, B. A., & Mayfield, S. P. (2015). Photosynthetic biomanufacturing in green algae; production of recombinant proteins for industrial, nutritional, and medical uses. *Photosynthesis Research, 123*(3), 227–239.

Raven, N., Rasche, S., Kuehn, C., Anderlei, T., Klöckner, W., Schuster, F., & Schillberg, S. (2015). Scaled-up manufacturing of recombinant antibodies produced by plant cells in a 200-L orbitally-shaken disposable bioreactor. *Biotechnology and Bioengineering, 112*(2), 308–321.

Richter, L. J., Thanavala, Y., Arntzen, C. J., & Mason, H. S. (2000). Production of hepatitis B surface antigen in transgenic plants for oral immunization. *Nature Biotechnology, 18*(11), 1167–1171.

Rouwendal, G. J., Wuhrer, M., Florack, D. E., Koeleman, C. A., Deelder, A. M., Bakker, H., et al. (2007). Efficient introduction of a bisecting GlcNAc residue in tobacco N-glycans by expression of the gene encoding human N-acetylglucosaminyltransferase III. *Glycobiology, 17*(3), 334–344.

Sainsbury, F., Sack, M., Stadlmann, J., Quendler, H., Fischer, R., & Lomonossoff, G. P. (2010). Rapid transient production in plants by replicating and non-replicating vectors yields high-quality functional anti-HIV antibodies. *PLoS One, 5*(11), e13976.

Sathish, K., Sriraman, R., Subramanian, B. M., Rao, N. H., Kasa, B., Donikeni, J., & Srinivasan, V. A. (2012). Plant expressed coccidial antigens as potential vaccine candidates in protecting chicken against coccidiosis. *Vaccine, 30*(30), 4460–4464.

Schouten, A., Roosien, J., van Engelen, F. A., Zilverentant, J. F., Bosch, D., Stiekema, W. J., & Bakker, J. (1996). The C-terminal KDEL sequence increases the expression level of a single-chain antibody designed to be targeted to both the cytosol and the secretory pathway in transgenic tobacco. *Plant Molecular Biology, 30*(4), 781–793.

Schroda, M. (2004). The Chlamydomonas genome reveals its secrets: Chaperone genes and the potential roles of their gene products in the chloroplast. *Photosynthesis Research, 82*(3), 221–240.

Shaaltiel, Y., Bartfeld, D., Hashmueli, S., Baum, G., Brill-Almon, E., Galili, G., & Aviezer, D. (2007). Production of glucocerebrosidase with terminal mannose glycans for enzyme replacement therapy of Gaucher's disease using a plant cell system. *Plant Biotechnology Journal, 5*(5), 579–590.

Sharma, M. K., Singh, N. K., Jani, D., Sisodia, R., Thungapathra, M., Gautam, J. K., Sharma, A. K. (2008). Expression of toxin co-regulated pilus subunit A (TCPA) of *Vibrio cholerae* and its immunogenic epitopes fused to cholera toxin B subunit in transgenic tomato (*Solanum lycopersicum*). *Plant Cell Reports, 27*(2), 307–318.

Sheshukova, E. V., Komarova, T. V., & Dorokhov, Y. L. (2016). Plant factories for the production of monoclonal antibodies. *Biochemistry (Moscow), 81*(10), 1118–1135.

Spiegel, H., Stöger, E., Twyman, R. M., & Buyel, J. F. (2018). Current status and perspectives of the molecular farming landscape. *Molecular Pharming: Applications, Challenges and Emerging Areas*, ed. A. R. Kermode (Hoboken, NJ: John Wiley & Sons, Inc.),3–23.

Stoger, E., Fischer, R., Moloney, M., & Ma, J. K. C. (2014). Plant molecular pharming for the treatment of chronic and infectious diseases. *Annual Review of Plant Biology, 65*, 743–768.

Tacket, C. O., Pasetti, M. F., Edelman, R., Howard, J. A., & Streatfield, S. (2004). Immunogenicity of recombinant LT-B delivered orally to humans in transgenic corn. *Vaccine, 22*(31–32), 4385–4389.

Takeyama, N., Yuki, Y., Tokuhara, D., Oroku, K., Mejima, M., Kurokawa, S., Kiyono, H. (2015). Oral rice-based vaccine induces passive and active immunity against enterotoxigenic *E. coli*-mediated diarrhea in pigs. *Vaccine, 33*(39), 5204–5211.

Thanavala, Y., Mahoney, M., Pal, S., Scott, A., Richter, L., Natarajan, N., & Mason, H. S. (2005). Immunogenicity in humans of an edible vaccine for hepatitis B. *Proceedings of the National Academy of Sciences, 102*(9), 3378–3382.

Tusé, D., Ku, N., Bendandi, M., Becerra, C., Collins, R., Langford, N., & Butler-Ransohoff, J. E. (2015). Clinical safety and immunogenicity of tumor-targeted, plant-made id-KLH conjugate vaccines for follicular lymphoma. *BioMed Research International, 2015*, 648143. doi: 10.1155/2015/648143

Twyman, R. M., Stoger, E., Schillberg, S., Christou, P., & Fischer, R. (2003). Molecular farming in plants: host systems and expression technology. *Trends in Biotechnology, 21*, 570–578.

Valenzuela, P., Medina, A., Rutter, W. J., Ammerer, G., & Hall, B. D. (1982). Synthesis and assembly of hepatitis B virus surface antigen particles in yeast. *Nature, 298*(5872), 347–350.

Vamvaka, E., Twyman, R. M., Murad, A. M., Melnik, S., Teh, A. Y. H., Arcalis, E., & Capell, T. (2016). Rice endosperm produces an underglycosylated and potent form of the HIV-neutralizing monoclonal antibody 2G12. *Plant Biotechnology Journal, 14*(1), 97–108.

Van Giap, D., Jung, J. W., & Kim, N. S. (2019). Production of functional recombinant cyclic citrullinated peptide monoclonal antibody in transgenic rice cell suspension culture. *Transgenic Research, 28*(2), 177–188.

Vetter, V., Denizer, G., Friedland, L. R., Krishnan, J., & Shapiro, M. (2018). Understanding modern-day vaccines: What you need to know. *Annals of Medicine, 50*(2), 110–120.

Ward, B. J., Makarkov, A., Séguin, A., Pillet, S., Trépanier, S., Dhaliwall, J., & Landry, N. (2020). Efficacy, immunogenicity, and safety of a plant-derived, quadrivalent, virus-like particle influenza vaccine in adults (18–64 years) and older adults ($\geq$ 65 years): Two multicentre, randomized phase 3 trials. *Lancet, 396*(10261), 1491–1503.

Webster, D. E., & Thomas, M. C. (2012). Post-translational modification of plant-made foreign proteins; glycosylation and beyond. *Biotechnology Advances, 30*(2), 410–418.

Wongsamuth, R., & Doran, P. M. (1997). Production of monoclonal antibodies by tobacco hairy roots. *Biotechnology and Bioengineering, 54*(5), 401–415.

Xu, X., Li, G. D., Kong, Y. Y., Yang, H. L., Zhang, Z. C., Cao, H. T., & Wang, Y. (1994). A modified hepatitis B virus surface antigen with the receptor binding site for hepatocytes at its C terminus: Expression, antigenicity, and immunogenicity. *The Journal of General Virology, 75*, 3673–3677.

Xu, J., Tan, L., Goodrum, K. J., & Kieliszewski, M. J. (2007). High-yields and extended serum half-life of human interferon α2b expressed in tobacco cells as arabinogalactan-protein fusions. *Biotechnology and Bioengineering, 97*(5), 997–1008.

Yang, Z., Drew, D. P., Jørgensen, B., Mandel, U., Bach, S. S., Ulvskov, P., & Petersen, B. L. (2012). Engineering mammalian mucin-type O-glycosylation in plants. *The Journal of Biological Chemistry, 287*(15), 11911–11923.

Yuki, Y., Nojima, M., Hosono, O., Tanaka, H., Kimura, Y., Satoh, T., Kiyono, H. (2021). Oral MucoRice-CTB vaccine for safety and microbiota-dependent immunogenicity in humans: a phase 1 randomised trial. *The Lancet Microbe, 2*(9), e429–e440.

Yusibov, V., Kushnir, N., & Streatfield, S. J. (2016). Antibody production in plants and green algae. *Annual Review of Plant Biology, 67*, 669–701.

Zhang, X., Buehner, N. A., Hutson, A. M., Estes, M. K., & Mason, H. S. (2006). Tomato is a highly effective vehicle for expression and oral immunization with Norwalk virus capsid protein. *Plant Biotechnology Journal, 4*(4), 419–432.

Chapter 5
Limitations, Biosafety, Ethics, Regulatory Issues in Molecular Farming in Plants

Molecular farming is a research-based application of biotechnology that uses genetically engineered crops to produce proteins and chemicals for pharmaceutical and other commercial reasons. It has gained popularity over the last two decades due to its potential benefits. The terms molecular farming, bio-farming, molecular pharming, phyto-manufacturing, plant pharma, plant biofactory, pharmaceutical gardening, and photo manufacturing are interchangeable (Basaran & Rodriguez-Cerezo, 2008).

This approach is a plant-based genetic transformation that may be carried out using stable and unstable gene transfer methods, such as gene transfer to nuclei, chloroplasts, and unstable gene transfer methods like viral vectors. The high expenses of medical treatments arising from the existing method are out of reach for most developing countries. Growing high demand for biomedicines has usually coincided with expensive rates and inefficient production systems (microbial, insect cells, transgenic animals, etc.). Transgenic crops as a new generation of bioreactors have received a lot of interest because of their advantages, such as the safety of recombinant proteins (growth factors, vaccines, enzymes, antibiotics, so on) and their potential for low-cost and large-scale production. All transgenic plants intended for molecular farming, like contemporary genetically modified (GM) plants, must undergo a rigorous health and environmental risk assessment before being employed for recombinant protein expression. Molecular farming plants, like all genetically modified plants, before field testing, commercialization, and release should be scrutinized routinely to verify that they don't create any harm to human health and the environment. Risk management is an attractive and difficult task that involves several disciplines, including ecology, agronomy, and molecular biology. It is primarily concerned with food and environmental safety (Ahmad et al., 2012; Chassy, 2010). Though, in addition to the risk management framework for transgenic crops used as feed, food, and processing (FFPs), plant molecular farming poses new challenges that may necessitate further biosafety considerations because of the nature of the recombinant genes used. In this regard, plant molecular farming also raises some

K. I. Wani, T. Aftab, *Plant Molecular Farming*, SpringerBriefs in Plant Science,
https://doi.org/10.1007/978-3-031-12794-6_5

new challenges in particular regarding aspects such as transgene amplification, diffusion, and the toxicity of recombinant protein accumulation that could trigger a need for biosafety concerns in the contamination of the food chain, feed chains, and environment (Ahmad, 2014).

1 Limitations, Biosafety, and Regulatory Issues of Molecular Forming

There are many limitations associated with molecular forming. One of which involves human health risks due to the allergens present in plants such as corn, which contains an industrial enzyme 'trypsin' and causes allergies. MF also impacts the environment, and the most severe impact is the mixing of the biopharma crops with the crops grown for human use using cross-pollination, leading to various severe risks to humankind.

Cross-pollination between corn plants is likely when crops are close together, as reported by researchers at the University of Maine. Extensive testing of temporal, physical, and biological barriers to prevent unintended gene transfers has revealed little danger. This risk quickly vanishes as the distance between crops grows. Downwind, traditional corn was planted 100 feet distant from GM corn to demonstrate the worst-case scenario. When hybrid corn was cultivated 100 feet downwind from GM corn, the results indicated a 1% possibility of cross-pollination. So the agencies are required to produce biopharmaceutical crops one mile apart from other crops under new biopharmaceutical laws (Seto, 2003).

Regulatory issues and biosafety are becoming increasingly important in molecular farming in plants or plant-derived tissue cultures/cells since one deals with genetically modified organisms (Eastham & Sweet, 2002). Public concern represents one of the most challenging concerns about using genetically engineered crops. Lack of communication among the agencies in charge of research, biosafety, and commerce has hampered progress in biopharming (Ramessar et al., 2008). The critical goals of biosafety rules and standards are risk management and eventually appropriate exchange of genetically modified organisms. The purposeful release of GMOs into the environment or the market, according to biosafety rules and regulations, should be led by a system of science-based risk management procedures. It's generally done by combining hazard identification and characterization of all risk components linked to a new GM crop or its derivatives (Craig et al., 2008).

The possible adverse human health consequences of the introduction of transgenic crops by molecular farming include the development of resistance in targeted disease populations. Also, transgenic organisms will lead to non-target effects connected with their products outside the plant (Craig et al., 2008). Also, their physiologically active products are supposed to induce biological reactions in animals and humans (Spök, 2007; Spök et al., 2008). In addition to these, the other specific concerns along with their management strategies are:

1.1   Environmental Risk Management

Environmental risk management can give high confidence to minimize threats by evaluating "ideas, or risk hypotheses," that forecast the possibility of unacceptable damaging occurrences. Poor issue formulation can raise the environmental threat by enabling the collection of excessive data, which can delay or prevent the introduction of environmentally valuable goods. It's impossible to predict the long-term environmental impact of molecular agricultural goods. Contamination of the food chain with medications derived from plants is a serious problem. This might occur due to the genetic material of transgenic plants transferred to food crops (Ahmad, 2014). Risk management for GMOs must be performed in a case-by-case manner using scientific methodologies. Many GMOs with their own set of novel characteristics and transgenic genes have been created. As a result, risk assessment has become very intricate and confusing, necessitating the development of both general and particular criteria for various species and features. These recommendations should concentrate on specific types of living modified organisms, their intended uses, apparent dangers, specific receiving settings, long-lasting monitoring of living altered creatures released into the surroundings, and the interaction among the engagement of competent national authorities responsible for risk management in biological diversity protection and sustainable use. However, there is currently no worldwide agreement on certain GMOs, such as GMOs employed as biocontrol agents for pests and diseases, live animal immunizations, GM trees, GM bio-fertilizers, bioremediation agents, and other GMOs that must be developed in the future. Moreover, despite the availability of a wealth of science-based information relating to risk management, such as knowledge of GMOs gained in specific environments over the past few years through research, field trials, associated risk assessments, contained use, and commercial releases, there are many frequent barriers of information access and understanding of how existing information can be used to encourage risk assessment (Jouzani, 2012).

1.2   Waste Management

If molecular farming activities grow in size, the leftover waste residue on the growing site and in storage facilities may become an issue. Appropriate waste management procedures (consigned in standard operating procedures) should be followed to assure that the substance does not enter the animal and human food chain or have an environmental impact (Freese 2002). In any event, judgments on using leftover products as animal feed (like starch manufactured from transgenic potato tubers) must be taken on a case-by-case approach, based on molecular farmed material's nature or danger, as well as the planned ultimate use (Breyer et al., 2009).

1.3   Contamination of the Food/Feed Chain

The concern of the contamination of the food/feed chain is prevented by the utilization of non-food and non-feed crops such as Lemna, tobacco, Arabidopsis, microalgae, etc. (Cox et al., 2006; Tremblay et al., 2010). However, when it's necessary to employ food crops, as in the case of certain monoclonal antibodies or oral vaccinations, containment and tight physical agronomic confinement measures are used to prevent cross-contamination. The plants are grown in tiny, isolated regions away from other plants to minimize physical mingling, crop damage, and post-planting field cleaning (Ma et al., 2005; Spök et al., 2008). Growing at separate times to assure harvesting at distinct times from other food and feed crops (Spök, 2007).

1.4   Containment Glasshouse Facilities

Seeds and pollen from transgenic crops in isolated glasshouses must be carefully monitored to ensure that they do not get outdoors. The plants must be labeled appropriately. There should be no cross-pollination of transgenic crops. High-level quality control is required for transgenic crops, gene constructions, DNA sequences, and experimental outcomes. Growing plants in a greenhouse require the same precision as growing them in a laboratory. Insects and pollen should be kept out of the greenhouse with good design. A Biosafety committee should supervise the facilities, and the amount of confinement should be determined by the kind of transgenic crops. The greenhouses must have filtered and regulated ventilation systems, as well as sanitation and water output control. The importance of autoclaving water and soil contaminants before leaving the greenhouse cannot be underscored.

1.5   Containment Laboratory Facilities

The two stages of physical confinement are primary and secondary biological containment. People and the nearby laboratory environment are protected from infectious agents by primary containment. Primary containment can be achieved using reasonable microbiological procedures and safety equipment. The main barriers are safety devices such as bio-safety cabinets, safety centrifuge cups, and sealed containers. Personal protection equipment, like lab coats and gloves, can function as the principal shield between personnel and pathogenic organisms while working in biological safety cabinets. Secondary containment keeps infectious materials out of the environment outside of the laboratory. Excellent facility design and operational standards offer secondary containment. Workspaces segregated from disinfection facilities, public areas, specialized ventilation systems, airlocks, and hand-washing facilities are examples of secondary barriers. Laboratory techniques, safety equipment, and facility design are the three fundamental factors of biological containment. Workers must examine the dangers associated with their jobs and learn how to use biosafety standards properly to ensure minimal exposure.

Apart from following reasonable laboratory procedures, particular practices for handling DNA Recombinant materials must be followed. Every institution must carefully adhere to the code of practice. There must be emergency protocols to cope with any situation at a recombinant DNA research facility (Sharma et al., 2002).

1.6 Gene Transmission and Unexpected Exposures

The biosafety strategies used in plant farming for preventing various risk assessments (gene transmission and incidental exposures of plant bioreactors) include the use of confined, separated physical containment facilities (hydroponics, greenhouses, plant cell suspension cultures, and glasshouses) and physiological containment (self-pollinating species), chloroplast transformation, sexually incompatible crops with wild relatives, non-pollinating species, cytoplasmic male-sterile transgenic crops, and engineered parthenocarpy (Gidoni et al., 2008; Obembe et al., 2011; Salehi Jouzani, 2012; Schillberg et al., 2013 and Valkova et al., 2013). Another technique used to reduce the impact of gene propagation is transgenic reduction. This strategy does not avoid transgene movement from transgenic plants to non-transgenic crops or wild relatives, but it does reduce the effects of such gene flow (Gressel & Valverde, 2009). To avoid food chain contamination, strong regulations must be implemented, like geologically separating the transgenic plants, growing them in greenhouses rather than open areas, harvesting as well as handling the transgenic plants with different types of equipment, or disinfecting the equipment thoroughly if it is also used on food crops (Rigano & Walmsley, 2005). Plant chloroplast transformation and male-sterile characteristics can also be used to accomplish containment. Because the chloroplast genome is inherited from the mother, transgenic pollen dispersion is limited. Also, it is critical to label genetically modified foods so that consumers may make informed decisions based on their preferences.

1.7 Horizontal Gene Transmission

The chance of horizontal transmission of genes from plants to microbes, especially when antibiotic resistance genes are used, is regarded to be exceedingly unlikely, given there has been no record of such an occurrence so far. Furthermore, it is healthy knowledge that plants, especially food plants, naturally host many bacteria carrying antibacterial drugs (Nielsen et al., 1998).

It ought to be emphasized that the majority of all the above-mentioned bio-safety techniques were prompted by the regulatory rules and guidelines of the many nations participating in plant molecular farming research, development, and industrial applications. However, several other concerns associated with bio-safety, risk assessment, and environmental impact are related to the release and large-scale growth of transgenic crops, as well as the safety of plant-derived products themselves.

It has been acknowledged that regulation implementation is still leaking since it is hard to track all phases of plant-derived product development and handling. As a result, there is also some negligence in terms of adherence on the part of the producers that goes unnoticed and untreated by regulatory agencies.

Regulatory authorities are confronted with a slew of issues regarding transgenic crop regulation. Because every biopharming product and the host system are distinct, every instance must have been addressed individually. The European Parliament and the Council of the European Union have allowed a 0.5 percent presence of transgenic materials in non-transgenic food when genetic materials are unavoidable and their advantages exceed their disadvantages (European Parliament, 2003). Recent efforts have been undertaken to contain transgenics and decrease environmental exposure (Chen et al., 2005; Ma et al., 2003). These constraints are still in their initial stages, and therefore more effort is required to address issues with molecular agricultural product regulations. As a result, higher conformity with good manufacturing practice standards is required throughout the planning process to assure the production of plant-derived goods.

2   Alternative Approaches Used to Reduce the Potential Risk of Plant Molecular Farming

Many alternative approaches (physical or biological) are available to reduce the contamination of food/feed chains or the environmental effect of plant molecular farming. Physical and geographical containment of GM plants, physical confinement methods, selective gene transcription, and also the invention of transcriptional methods are among the most effective (Martine et al., 2009).

2.1   Physical Containment

Planting genetically modified plants in physical structures is one example of the broader set of precautions that might minimize environmental contamination. Plants produced in enclosed structures are tobacco, potatoes, and green crops including lettuce, spinach, and alfalfa. Plastic tunnels, greenhouse production facilities, labs, and growth facilities such as phytotrons are examples of the physical enclosure. In several nations, e.g., US and Canada, large-scale subsurface facilities have also been employed (Tackaberry et al., 2003). On the other hand, physical containment has significant drawbacks, like the increasing economic resources needed to cultivate plants in confinement (Martine et al., 2009).

2.2   Spatial Containment

Spatial separation comprises various approaches to reduce fertilization among pharmaceutical and commercial crops. These may be especially important when pharma/industrial cultivation of GM plants can be done in a short space.

2.3   Contained Field Trials

Once a plant is allowed for small-scale research or, subsequently, industrial commercialization, it must undergo extensive risk assessment. At this point, the scientists involved in the national or international regulatory authorities and the biosafety committee should decide whether it is permissible to release the specific genetically modified crop and, if so, what limits should be placed on them. To minimize the potential environmental effect of the release experiment, field confinement must be in place. Sexually compatible species, blooming prevention, the adoption of subsequent monitoring, and male-sterile line techniques are all examples.

Data needed for risk management comprises; (i) Basic information, (ii) DNA donor, receiving species, and genetic crop, (iii) environment and release conditions, (iv) genetic transformation interactions, and (v) regulation, observation, and waste management.

While moving forward with outdoor experiments, it is essential to obtain safety data in a controlled condition. A glasshouse, playhouse, or greenhouse might create a climate that regulates light, humidity, ventilation, temperature, and practical barriers in preventing microbiological organisms or insects from entering and efficiently trapping pollen grains and biological materials. Information on the factors mentioned above must be prepared consistently for regulatory authorities to make an informed decision about the correct evaluation of risks associated with the usage of transgenic crops before allowing their distribution on a large scale.

2.4   Targeted Expression

Another technique for reducing accidental exposure to a pharmacological product is to target the expression of the development of concern to a few unique plant sections. Particular promoters with tissue-specific expression are used to accomplish this. One can apply numerous solutions based on the potential benefits and drawbacks of enhanced yields, purification process simplicity, and biosafety concerns.

2.5   Transient Expression

Organizations wishing to develop vaccinations or other pharmaceuticals on a medium and small scale are increasingly employing this method (Vézina et al., 2009). Increasing amounts of protein synthesis can be reached for a short space of time. Therefore, in this regard, the gene is readily available in plant tissues for a limited timeframe and cannot be passed down to future generations. Another advantage of the approach is that it is usually used in a controlled environment. Direct delivery techniques (e.g., particulate bombarding or microinjection), and bacterial infection can all be employed to generate a transitory expression of DNA or RNA (Gils et al., 2005). Agro infiltration, which involves inoculating leaves using recombinant *Agrobacterium tumefaciens*, is one of the most promising transient expression techniques (D'Aoust et al., 2009; Marillonnet et al., 2005).

Several other crop confinement approaches may also effectively reduce or eliminate gene flow. Apomixis, genetic incompatibility, cleistogamy, temporal and tissue-specific regulation, and transgenic mitigation, are just a few examples (Daniell, 2002). It's vital to remember that most of the confinement techniques listed above aren't ready for commercial utilization (Ellstrand, 2003). Furthermore, some others do not stop genetic transformation caused by seed dispersal after cultivation, harvesting, and transportation. As a result, biological barriers are unlikely to restrict genetic drift completely. Applying one or more of these methods on a case-by-case approach, in cooperation with certain other confinement approaches, and taking into account the inherent qualities of each particular crop may also be required to provide a high level of environmental management.

2.5.1   Transient Expression via Virus Infection

Viruses potentially transmit genetic material linked with the coat protein-coding genotype into plants, resulting in rapid genetic transformation due to RNA viruses' ability to reproduce large numbers of sequences in the host. This approach's gene expression system was created using Tombusvirus (tomato bushy stunt virus), Tobamovirus (tobacco mosaic virus), and comovirus (cowpea mosaic virus) which resulted in an increased expression without integrating a transgene (Cañizares et al., 2005; Porta & Lomonossoff, 2002). The drawbacks of this are the recombinant protein's instability and the need for early tissue processing to prevent degradation. A pharmaceutical business used this strategy to research to develop idiotype vaccines to treat B-cell non-Hodgkin's lymphoma (McCormick et al., 2008).

2.5.2   Agroinfiltration Method

Agroinfiltration involves penetration of bacteria inside the leaf tissue, which results in t-DNA transmission to a large number of cells. t-DNA doesn't integrate with genes and generates temporarily at a greater level, resulting in more excellent nucleoprotein production (Kapila et al., 1997). It is a rapid and productive approach, and it also has the advantage of allowing the construction of multi-subunit proteins to be studied *in vitro* propagation due to the presence of several chromosomes in a single tissue (Vaquero et al., 1999; Vézina et al., 2009).

2.5.3   Magnifection Technology

Due to the difficulties of pathogen and Agrobacterium-mediated transgenic protein production in plants to provide upregulation of numerous peptides needed for multimeric protein assembly, a new and robust technology known as magnifection was developed. This approach requires infecting the host plant using Agrobacterium to transfer bacteriophages to the plant body. Magnifection resulted in increased infectivity and amplification, allowing for the simultaneous production of several proteins that can assemble into a functional multimeric protein. Flagellin F1–V fusion protein, hepatitis B core antigen (HBcAg), and IgG has been successfully tested by practicing this approach.

2.6   Chloroplast Transformation

Because of the special characteristic of definitive integration of transgenes into the plasmid via general recombination, the present approach provides an adequate expression system, as site-specific mutation reduces the issue of gene silencing which can cause variations in transcription across various transgenic plants (Ruhlman et al., 2010). Other advantages include transgene confinement due to the chloroplast's prokaryotic origin maternal inheritance found in plastids and multigene engineering potential (Kumar et al., 2012). Because each cell has several chloroplasts, each chloroplast carries countless copies of the genome; this system may achieve a very high degree of recombinant protein production. As chloroplasts are maternally inherited; therefore, transfer to other plants through pollen is avoided.

3   Ethical Issues

Science has significantly influenced human life, offering various breakthroughs that have improved many lives. Researchers are widely regarded as reliable and ethical for agricultural research and their function in the production of food is essentially desirable. Although it is widely acknowledged that new technologies challenge current values and institutions and drive change in conventional views of nature and human identity, this viewpoint has been influenced by the arrival of genetic engineering.

3.1 Ethical Implications of Hype

Expanded therapeutic assertions should be seen as a hallmark of bioengineering and have also been highlighted in biopharmaceutical development (Bloomfield & Doolin, 2011; Brown, 2003). Grant, a researcher in the human science of aspirations, has stressed the importance of establishing future achievement desires in mobilizing support for technoscience (Borup et al., 2006). The history of biopharming presents moral problems concerning the morality of promoting current technoscience through ineffectively founded assertions, but not exceptionally so. As Brown (2003) has suggested in conjunction with heterologous transplant research, implausible advantage assertions can be balanced with hazards, allowing, e.g., a level of creature endurance that would not be otherwise permitted. Therefore, beneficial statements (e.g., molecular farming would permit the production of therapeutics for disorders that are now untreatable or that it will sustain positive, low-cost medication to underdeveloped nations) should be given more weight. It also implies that the morality of expressing such claims is taken into account. This consistency of beneficial assertions too is important for determining whether it is moral to avoid biopharming in any of its types (Taussig, 2004).

3.2 Inflicting or Exposing to Harm

Individuals could acquire diseases spread by biopharma substances, despite pharmaceutical regulators' best efforts to prevent it. New pollutants which haven't been cleansed by refining methods may also impact individuals. It might necessitate a debate regarding whether or not patients should be told about the medications' origins (Rehbinder et al., 2009). Open-air biopharming seems to have the potential to affect the surroundings, particularly living animals that come into contact with biopharmaceutical plants, creatures, or buildups. This may include humans, whether agricultural workers or who ingest food infected by open-air biopharming, as well as other animals, who may be affected. It thus requires a careful depiction of damage or potential damage when there is none previously existing, hence necessitating justification (Taussig, 2004).

3.3 Cost-Benefit Analysis: The Importance of Alternatives and Efficiency Improvements

A classic legal strategy assesses the challenges and possible risks against the potential benefits and advantages. It brings us back to the constant nature of benefit claims and the issue of possibilities. Molecular farming is in demand throughout a range of "platforms" at the same time. Given this and the instabilities that still define the invention, it would be challenging to argue that there is no choice between stages. More than a specialized possibility, financial and intellectual-property consider- ations appear to determine stage selection (Fischer et al., 2012). Such modifiability is almost as important as prospective advantages when it comes to moral debates concerning the acceptability of causing harm through the use of biopharming. The need for fetched investment monies has become such a major catalyst of biopharm- ing, that monetary advantages (by taking a toll on investment funds) could over- come the inevitable burden.

3.4   Hubris, Irresponsibility, and Wisdom

Biopharming, or at least some of its forms, may also be regarded to exemplify a hubristic and reckless intrusion into life, similar to other physical interventions characterized by a wide range of security vulnerabilities and cluelessness, as well as the possibility of causing irreparable harm to the humans and the environment (Fiester, 2008). Biopharming (too uninterestingly) may generate worries that we are doing it "just because we can" rather than because it is necessary or prudent. This, in turn, raises concern about the morality and direction of our technological framework.

4   Conclusion

In conclusion, molecular farming in plants aims to offer a cheaper and safer source of biopharmaceuticals, and to date, several pharmaceutical proteins have been expressed in plants. Nowadays, with the scientific advances in biotechnology, gene transfer methods in plants have considerably developed. Compared with other microbial and animal expression systems, these transgenic plants have various advantages in easy production, cost, safety, etc., for producing pharmaceutical bio- molecules. So far, lots of valuable pharmaceutical proteins and antibodies have been made with the help of this method, which remarkably has helped the treatment of patients, especially in developing countries where the production and preservation costs of such medicines cannot be afforded. However, there are several limitations in molecular farming. Currently, much more effort is devoted to overcoming these limitations. Recently, technological breakthroughs have enhanced gene transfer, recombinant protein synthesis, and purification technologies, allowing scientists to modify plants and utilize them as bioreactors for the mass manufacturing of various medications.

Furthermore, different bio-safety and environmental issues have been examined for molecular farming plants. To promote public acceptance, biosecurity, clinical, and commercialization investigations of plant molecular farming, scientific and regulatory risk assessment, and risk management methodologies and standards must be developed. Efforts are being devoted to overcoming the challenges of plant molecular farming, and it is hoped that shortly it will witness outstanding achievements. However, a lot more has to be done to make molecular farming products a success.

References

Ahmad, K. (2014). Molecular farming: Strategies, expression systems, and bio-safety considerations. *Czech Journal of Genetics and Plant Breeding, 50*(1), 1–10.

Ahmad, P., Ashraf, M., Younis, M., Hu, X., Kumar, A., Akram, N. A., & Al-Qurainy, F. (2012). Role of transgenic plants in agriculture and biopharming. *Biotechnology Advances, 30*(3), 524–540. https://doi.org/10.1016/j.biotechadv.2011.09.006

Basaran, P., & Rodriguez-Cerezo, E. (2008). An assessment of emerging molecular farming activities based on patent analysis (2002–2006). *Biotech Bioprocess Engineer, 13*(3), 304–12.

Bloomfield, B. P., & Doolin, B. (2011). Imagination and technoscientific innovations: Governance of transgenic cows in New Zealand. *Social Studies of Science, 41*(1), 59–83.

Borup, M., Brown, N., Konrad, K., & Van Lente, H. (2006). The sociology of expectations in science and technology. *Technology Analysis & Strategic Management, 18*(3–4), 285–298.

Breyer, D., Goossens, M., Herman, P., & Sneyers, M. (2009). Biosafety considerations associated with molecular farming in genetically modified plants. *Journal of Medicinal Plants Research, 3*(11), 825–838.

Brown, N. (2003). Hope against hype-accountability in biopasts, presents and futures. *Science and Technology Studies, 16*(2), 3–21.

Cañizares, M. C., Nicholson, L., & Lomonossoff, G. P. (2005). Use of viral vectors for vaccine production in plants. *Immunology and Cell Biology, 83*(3), 263–270.

Chassy, B. M. (2010). Food safety risks and consumer health. *New Biotechnology, 27*(5), 534–544.

Chen, M., Liu, X., Wang, Z., Song, J., Qi, Q., & Wang, P. G. (2005). Modification of plant N-glycans processing: The future of producing therapeutic protein by transgenic plants. *Medicinal Research Reviews, 25*(3), 343–360.

Cox, K. M., Sterling, J. D., Regan, J. T., Gasdaska, J. R., Frantz, K. K., Peele, C. G., et al. (2006). Glycan optimization of a human monoclonal antibody in the aquatic plant Lemna minor. *Nature Biotechnology, 24*(12), 1591–1597.

Craig, W., Tepfer, M., Degrassi, G., & Ripandelli, D. (2008). An overview of general features of risk assessments of genetically modified crops. *Euphytica, 164*, 853–880.

Daniell, H. (2002). Molecular strategies for gene containment in transgenic crops. *Nature Biotechnology, 20*(6), 581–586.

D'Aoust, M. A., Lavoie, P. O., Belles-Isles, J., Bechtold, N., Martel, M., & Vézina, L. P. (2009). Transient expression of antibodies in plants using syringe agroinfiltration. In *Recombinant proteins from plants* (pp. 41–50). Humana Press.

Eastham, K., & Sweet, J. (2002). *Genetically modified organisms (GMOs): The significance of gene flow through pollen transfer* (pp. 1–74). European Environment Agency.

Ellstrand, N. C. (2003). Going to "great lengths" to prevent the escape of genes that produce specialty chemicals. *Plant Physiology, 132*(4), 1770–1774.

European Parliament. (2003). Regulation (EC) no 1829/2003 of the European Parliament and of the council of 22 September 2003 on genetically modified food and feed. *Official Journal of the European Union, L268*, 1–23.

Fiester, A. (2008). Justifying a presumption of restraint in animal biotechnology research. *The American Journal of Bioethics, 8*(6), 36–44.

Fischer, R., Schillberg, S., Hellwig, S., Twyman, R. M., & Drossard, J. (2012). GMP issues for recombinant plant-derived pharmaceutical proteins. *Biotechnology Advances, 30*(2), 434–439.

Freese B. (2002). Manufacturing drugs and chemical crops:Biopharming poses new threats to consumers, farmers,food companies and the environment. Available from GE Food Alert, www.gefoodalert.org.

Gidoni, D., Srivastava, V., & Carmi, N. (2008). Site-specific excisional recombination strategies for elimination of undesirable transgenes from crop plants. *In Vitro Cellular & Developmental Biology - Plant, 44*(6), 457–467.

Gils, M., Kandzia, R., Marillonnet, S., Klimyuk, V., & Gleba, Y. (2005). High-yield production of authentic human growth hormone using a plant virus-based expression system. *Plant Biotechnology Journal, 3*(6), 613–620.

Gressel, J., & Valverde, B. E. (2009). A strategy to provide long-term control of weedy rice while mitigating herbicide resistance transgene flow, and its potential use for other crops with related weeds. *Pest Management Science (Formerly Pesticide Science), 65*(7), 723–731.

Jouzani, G. S. (2012). Risk assessment of GM crops; challenges in regulations and science. *BioSafety, 1*, e113.

Kapila, J., De Rycke, R., Van Montagu, M., & Angenon, G. (1997). An agrobacterium-mediated transient gene expression system for intact leaves. *Plant Science, 122*(1), 101–108.

Kumar, S., Hahn, F. M., Baidoo, E., Kahlon, T. S., Wood, D. F., McMahan, C. M., et al. (2012). Remodeling the isoprenoid pathway in tobacco by expressing the cytoplasmic mevalonate pathway in chloroplasts. *Metabolic Engineering, 14*(1), 19–28.

Ma, J. K., Drake, P. M., & Christou, P. (2003). The production of recombinant pharmaceutical proteins in plants. *Nature Reviews. Genetics, 4*(10), 794–805.

Ma, J. K. C., Barros, E., Bock, R., Christou, P., Dale, P. J., Dix, P. J., & Twyman, R. M. (2005). Molecular farming for new drugs and vaccines: current perspectives on the production of pharmaceuticals in transgenic plants. *EMBO Reports, 6*(7), 593–599.

Marillonnet, S., Thoeringer, C., Kandzia, R., Klimyuk, V., & Gleba, Y. (2005). Systemic agrobacterium tumefaciens–mediated transfection of viral replicons for efficient transient expression in plants. *Nature Biotechnology, 23*(6), 718–723.

Martine, G., Philippe, H., & Myriam, S. (2009). Biosafety considerations associated with molecular farming in genetically modified plants. *Journal of Medicinal Plants Research, 3*(11), 825-838.

McCormick, A. A., Reddy, S., Reinl, S. J., Cameron, T. I., Czerwinkski, D. K., Vojdani, F., et al. (2008). Plant-produced idiotype vaccines for the treatment of non-Hodgkin's lymphoma: Safety and immunogenicity in phase I clinical study. *Proceedings of the National Academy of Sciences, 105*(29), 10131–10136.

Nielsen, K. M., Bones, A. M., Smalla, K., & van Elsas, J. D. (1998). Horizontal gene transfer from transgenic plants to terrestrial bacteria–is a rare event? *FEMS Microbiology Reviews, 22*(2), 79–103.

Obembe, O. O., Popoola, J. O., Leelavathi, S., & Reddy, S. V. (2011). Advances in plant molecular farming. *Biotechnology Advances, 29*(2), 210–222. https://doi.org/10.1016/j.biotechadv.2010.11.004

Porta, C., & Lomonossoff, G. P. (2002). Viruses as vectors for the expression of foreign sequences in plants. *Biotechnology & Genetic Engineering Reviews, 19*(1), 245–292.

Ramessar, K., Capell, T., Twyman, R. M., Quemada, H., & Christou, P. (2008). Trace and traceability—a call for regulatory harmony. *Nature Biotechnology, 26*(9), 975–978.

Rehbinder, E., Engelhard, M., Hagen, K., & Jørgensen, R. B. (2009). R. Pardo-Avellaneda, A. Schnieke et al. Promises and risks of biopharmaceuticals derived from genetically modified plants and animals.

Rigano, M. M., & Walmsley, A. M. (2005). Expression systems and developments in plant-made vaccines. *Immunology and Cell Biology, 83*(3), 271–277. https://doi.org/10.1111/j.1440-1711.2005.01336.x

Ruhlman, T., Verma, D., Samson, N., & Daniell, H. (2010). The role of heterologous chloroplast sequence elements in transgene integration and expression. *Plant Physiology, 152*(4), 2088–2104.

Salehi Jouzani, G. (2012). Risk assessment of GM crops; challenges in regulations and science. *Biosafety, 1*, e113.

Schillberg, S., Raven, N., Fischer, R., Twyman, R. M., & Schiermeyer, A. (2013). Molecular farming of pharmaceutical proteins using plant suspension cell and tissue cultures. *Current Pharmaceutical Design, 19*(31), 5531–5542. https://doi.org/10.2174/1381612811319310008

Seto, R. C. (2003). Selling the pharm: The risks, benefits, and regulation of biopharmaceuticals. *Environmental Law and Policy Journal, 27*, 443–466.

Sharma, K. K., Sharma, H. C., Seetharama, N., & Ortiz, R. (2002). Development and deployment of transgenic plants: Biosafety considerations. *In Vitro Cellular & Developmental Biology - Plant, 38*(2), 106–115.

Spök, A. (2007). Molecular farming on the rise–GMO regulators still walking a tightrope. *Trends in Biotechnology, 25*(2), 74–82.

Spök, A., Twyman, R. M., Fischer, R., Ma, J. K., & Sparrow, P. A. (2008). Evolution of a regulatory framework for pharmaceuticals derived from genetically modified plants. *Trends in Biotechnology, 26*(9), 506–517.

Tackaberry, E. S., Prior, F., Bell, M., Tocchi, M., Porter, S., Mehic, J., et al. (2003). Increased yield of heterologous viral glycoprotein in the seeds of homozygous transgenic tobacco plants cultivated underground. *Genome, 46*(3), 521–526.

Taussig, K. S. (2004). Bovine abominations: Genetic culture and politics in the Netherlands. *Cultural Anthropology, 19*(3), 305–336.

Tremblay, R., Wang, D., Jevnikar, A. M., & Ma, S. (2010). Tobacco, is a highly efficient green bioreactor for the production of therapeutic proteins. *Biotechnology Advances, 28*(2), 214–221.

Valkova, R., Apostolova, E., & Naimov, S. (2013). Plant molecular farming: Opportunities and challenges. *Journal of the Serbian Chemical Society, 78*, 407–415.

Vaquero, C., Sack, M., Chandler, J., Drossard, J., Schuster, F., Monecke, M., et al. (1999). Transient expression of a tumor-specific single-chain fragment and a chimeric antibody in tobacco leaves. *Proceedings of the National Academy of Sciences, 96*(20), 11128–11133.

Vézina, L. P., Faye, L., Lerouge, P., D'Aoust, M. A., Marquet-Blouin, E., Burel, C., & Gomord, V. (2009). Transient co-expression for fast and high-yield production of antibodies with human-like N-glycans in plants. *Plant Biotechnology Journal, 7*(5), 442–455.

Chapter 6
Conclusion and Perspectives

1 Conclusion

The exotic universe of plant molecular farming has developed into a versatile and dynamic technology in the recombinant protein market, particularly in the pharmaceutical sector. This emergence of the plant-based recombinant protein factories has been possible due to the development of high-end production facilities, the emergence of transient expression systems for high yield, improvement in downstream processing strategies, and targeting niche products. Moreover, the increasing number of clinical trials of plant-based pharmaceuticals will ensure that more and more recombinant plant-based drugs reach the market in near future. These plant-made pharmaceuticals comprise a variety of recombinant proteins including vaccines, antibodies, and replacement human proteins, with a focus on the oral delivery of these drugs. In addition to pharmaceutical proteins, plant molecular farming also remains an attractive and profitable venture for non-pharmaceutical proteins such as cosmetic ingredients and research-grade reagents.

Plants have both technical and economic advantages over conventionally used expression platforms for recombinant protein expression. The different expression strategies used in plant molecular farming such as nuclear expression, chloroplast expression, transient expression, and viral transfection have their own unique features which enable them to produce a diverse range of target proteins. Even though many technical and scientific challenges associated with plant molecular farming were met in recent years, the major barrier to the wide acceptance of plant-based therapeutic proteins is the associated regulatory burden. Due to great scalability, low costs, and fewer regulatory challenges associated with the plant-based production of non-pharmaceutical proteins, their commercialization is faster than recombinant pharmaceutical proteins. Despite having the potential for high-profit margins, the therapeutic proteins require extensive preclinical and clinical trials, with backing from industrial partners having expertise in drug development, which hinders

K. I. Wani, T. Aftab, *Plant Molecular Farming*, SpringerBriefs in Plant Science, https://doi.org/10.1007/978-3-031-12794-6_6

their commercialization. Some of the studies related to recombinant therapeutic protein production do not even go beyond expression, purification, and cell-based analysis. Progress along the value chain has other requirements as well which include toxicity studies, identification of biomarkers for patient stratification and therapeutic monitoring, clinical trials, acceptance by the health agencies, and approval by regulatory agencies. The regulatory framework and restrictions greatly influence the worldwide acceptance of plant molecular farming. Furthermore, the commercial considerations like intellectual property/freedom to operate portfolio, potential competitors, time to market, and market share also decide the commercial success of a particular plant-based therapeutic protein. The unveiling of more plant-based therapeutics in the market requires huge investment and mutual cooperation with clinicians, regulatory authorities, and pharmaceutical industries. These things only make sense if the resultant product produced in plants outshines the same product when produced in Chinese hamster ovary cells or microbes. Plant systems' economic feasibility and production capability, together with increasing demand for pharmaceutically and industrially useful proteins, suggest a bright future for plant-made biologics. The impact of plant molecular farming will increase as the platforms become standardized and optimized according to industry norms to meet the good manufacturing practice standards while still maintaining the economic, technical, and societal advantages that may come from the production of resultant pharmaceutical proteins.

2 Perspectives and Future Opportunities

Plant-based products currently under preclinical and clinical trials show that a single expression platform cannot dominate, rather each platform has its advantages and disadvantages, each with different variants suitable for the production of specific plant-made pharmaceutical products. The platform diversity may be interesting at the research and development level, but its commercial impact can be uncertain. Platform diversity provides more operational freedom and multiple licensing opportunities, however, the lack of a generic upstream and downstream infrastructure forms a major barrier, particularly for bio-manufacturing and pharmaceutical companies working with established expression systems.

Despite having many advantages, plant-based systems cannot compete with the already established systems based on microbes and mammalian cells in terms of good manufacturing practice and regulatory approval at an industrial level. The process of producing recombinant therapeutic proteins is slow as it takes a lot of time from the lab bench to its commercialization. So the commercial potential of this technology could also be exploited by the development of non-pharmaceutical diagnostic products, industrial enzymes, and veterinary vaccines as they have fewer regulatory barriers as compared to therapeutic proteins (Tschofen et al., 2016; Schillberg et al., 2019). It can also be used for manufacturing diagnostic reagents and rapid response vaccines against newly emerging infections like SARS-CoV-2.

In the near future, smaller companies may start producing new products via molecular farming, with a focus on niche markets such as emergency treatments, personalized medicines, low-cost vaccines, and orphan diseases using new approaches like plant virus-like particles (VLPs). The special features of plant systems may also be exploited by generating products that require minimum processing for mucosal delivery, which increases their longevity and efficacy.

The manufacturing capacity of existing molecular farming facilities could be boosted by using contract-manufacturing organizations. Moreover, the increase in the number of plant-made pharmaceuticals under preclinical and clinical trials will make the concerned regulatory authorities familiar with upstream and downstream procedures and how these can be transferred into good manufacturing practice-compliant operations.

Another promising step toward the rapid development of plant molecular farming is the development of genome editing tools like Clustered Regularly Interspaced Palindromic Repeats (CRISPR), CRISPR associated (Cas9) system, transcription activator-like effector nucleases (TALENs), and zinc finger nucleases. These tools simplify the knockout of unwanted functions and are better than RNA interference and random mutagenesis approaches used in the past. These genome editing tools will help in the development of superior plant lines and cell suspension cultures. These will be particularly useful in the production of plant-made pharmaceuticals with defined glycan structures to improve efficacy, safety, serum stability, and immunogenicity (in the case of vaccines. These genome editing tools could also be used for the generation of stable transgenic lines. The full potential of plant molecular farming for the production of cost-effective therapeutic proteins will be evident in the immediate future.

References

Schillberg, S., Raven, N., Spiegel, H., Rasche, S., & Buntru, M. (2019). Critical analysis of the commercial potential of plants for the production of recombinant proteins. *Front Plant Sci, 10*, 720.

Tschofen M, Knopp D, Hood E, Stöger E. (2016). Plant Molecular Farming: Much More than Medicines. *Annu Rev Anal Chem, 9*(1):271–94. doi: 10.1146/annurev-anchem-071015-041706.

Printed in the United States
by Baker & Taylor Publisher Services